ECONOMIC PROGRESS AND THE ENVIRONMENT

ECONOMIC PROGRESS AND THE ENVIRONMENT
One Developing Country's Policy Crisis

Douglas Southgate
and
Morris Whitaker

New York   Oxford
OXFORD UNIVERSITY PRESS
1994

TO MYRIAM AND MARGE

Oxford University Press

Oxford New York Toronto
Delhi Bombay Calcutta Madras Karachi
Kuala Lumpur Singapore Hong Kong Tokyo
Nairobi Dar es Salaam Cape Town
Melbourne Auckland Madrid

and associated companies in
Berlin Ibadan

Copyright © 1994 by Oxford University Press, Inc.

Published by Oxford University Press, Inc.,
200 Madison Avenue, New York, New York 10016

Oxford is a registered trademark of Oxford University Press

All rights reserved. No part of this publication may be reproduced,
stored in a retrieval system, or transmitted, in any form or by any means,
electronic, mechanical, photocopying, recording, or otherwise,
without the prior permission of Oxford University Press.

Library of Congress Cataloging-in-Publication Data
Southgate, Douglas DeWitt, 1952–
Economic progress and the environment : one developing country's
policy crisis / Douglas Southgate and Morris Whitaker.
p. cm. Includes bibliographical references and index.
ISBN 0-19-508786-0
1. Environmental policy—Ecuador.
2. Environmental degradation—Ecuador.
3. Natural resources—Ecuador—Management.
4. Sustainable development—Ecuador.
I. Whitaker, Morris D. II. Title.
HC204.5.E5S66 1994
363.7'009866—dc20   94–5615

2 4 6 8 9 7 5 3 1

Printed in the United States of America
on acid-free paper

Acknowledgments

The two authors have consulted with a number of people during the preparation of this volume. Helpful advice and reviews have been received from our colleagues at the Instituto de Estrategias Agropecuarias (IDEA): Neptalí Bonifaz (the executive director), Carlos Camacho, Duty Greene, Fernando Ortiz (who contributed to early drafts of the fifth and ninth chapters), Hugo Ramos, and Jorge Soria. Similar support was provided by current and former professional staff members of the Agriculture and Natural Resources Office of the U.S. Agency for International Development (AID) mission in Quito: David Alverson, Howard Clark, Fausto Maldonado, Richard Peters, Jack Rosholt, and Ronald Ruybal. David Edwards, Greg Lagana, John Savage, and David Seckler, all of the U.S. embassy in Quito, provided valuable input on the seventh, eighth, and ninth chapters. Rodolfo Barniol (of Aqualab, in Guayaquil, Ecuador), Kate Clark (of Quito, Ecuador), Paul Colinvaux (formerly of Ohio State University and now with the Smithsonian Tropical Research Institute, in Panama), David Daines (of Molinos Champion, in Guayaquil), Bruce Epler (of the University of Rhode Island), Michael Hanrahan (of Development Alternatives, Inc., in Bethesda, Maryland), Jorge Jimenez (of Quito, Ecuador), Bruce Kernan (of Quito, Ecuador), David Lee (of Cornell University), Norman Myers (of Oxford, England), and John Perez-García (of the University of Washington) also commented on draft chapters. Robert Costanza (of the University of Maryland) and Richard Rice (of Conservation International) provided helpful comments on the tenth and eleventh chapters at the request of Oxford University Press.

This book could not have been written had we not been able to carry out research in Ecuador for an extended period. This opportunity was provided by the Quito AID mission and IDEA, where the environment for independent study of policy issues is excellent. One of IDEA's most important assets is its excellent research assistants: Paul Arellano, María Arguello, Francisco Barba, Manuel Bonifaz, Marc Carey, Lisa Chase, Marc Coles-Ritchie, Rubén Flores, Doris Ortiz, Karin Perkins, Paul Sabin, and Pablo Salazar. IDEA also contracted with Guillermo Meza and Kate Clark for map preparation.

The current and former leadership of the AID mission and also the embassy merits acknowledgment. Research for this volume began when Franz Almaguer was the AID director and continued under the leadership of Charles Costello, and John Sanbrailo. Like them, Ambassador Paul Lambert and Deputy Chief of Mission James Mack were supportive of our work.

Dr. Southgate's secretary at Ohio State University, Jen Heller, helped track down a lot of information and literature citations in the United States. She and Dr. Whitaker's secretary, True Rubal, deserve our heartfelt thanks. So does Mary Sue Southgate, who located data and citations.

Above all, we would like to thank our wives and children, who put up with the frequent late hours and international travel required during the preparation of this volume. Thank you, Myriam and Marge.

Needless to say, all errors and omissions in the book are our responsibility and all opinions expressed in the pages that follow are ours alone and do not necessarily represent positions taken by any of the aforementioned institutions.

Contents

I The Causes of Renewable Resource Degradation

1. Environmental Crisis in the Latin American Countryside   3
   Outline of the Book   5
   The Study's Geographic Focus   8

2. Causes of Increasing Resource Scarcity   10
   Population Growth   11
   Internal Migration   13
   Income Growth   15
   Increasing Domestic Demand for Agricultural Commodities   16
   Macroeconomic and Sectoral Policies and Performance of the Agricultural Economy   17
   The Challenge of Increasing Scarcity   20

3. Policy Crisis and Environmental Degradation   21
   Discriminatory Macroeconomic and Sectoral Policies   22
   Factor Market Distortions   23
   Failure to Invest in the Rural Economy's Scientific Base   27
   Labor Markets and the Environment   28
   An Ideal Set of Policies for Environmental Degradation   29

II Case Studies

4. Tropical Deforestation   33
   Historical Trends   34
   Current Magnitude   36
   Net Returns   37
   Causes of Agricultural Colonization   42
   Summary and Conclusions   45

5. Farmland Degradation   48
   Soil Erosion and Its Origins   48
   On-Farm Consequences of Soil Loss   51
   Why Farmers Allow Their Land to Deteriorate   51
   Land Degradation and Declining Commodity Prices   53
   New Directions in Soil Conservation Policy   56

6. Waste and Misallocation of Water Resources   57
   Water Resources for Agriculture   58
   The Water Law and INERHI   59
   Irrigation Projects   61
   Hydroelectricity Development   68
   Potable Water Systems   72
   Water Resource Degradation   73
   Summary and Conclusions   77

7. Oil Industry Pollution in the Ecuadorian Amazon   79
   Environmental Impacts   80
   Current Regulatory Environment   82
   The International Controversy Surrounding Development of Block 16   83
   The Economics of Pollution Control   85
   Summary and Conclusions   89

8. Shrimp Mariculture and Coastal Ecosystems   90
   Alteration of Coastal Ecosystems   91
   Consequences of Coastal Ecosystem Disturbance   94
   Causes of Environmental Degradation along the Ecuadorian Coast   95
   The Future of Ecuadorian Mariculture and the Country's Coastal Ecosystems   98

9. Tourism and Species Preservation in the Galápagos   101
   A Short History of the Galápagos   102
   Conservation Initiatives   104
   The Growth of Tourism   105
   Tourism Pricing Issues   106
   Conclusions   109

III Conclusions and Recommendations

10. Development and the Environment:
    Some Common Fallacies   *113*
    Is Population Growth Out of Control in Ecuador?   114
    *Will Population Growth Outstrip What the Environment
       Can Support?   115*
    Must Economic Growth Always Harm the Environment?   117
    *Is Economic Liberalization Inherently Unfair to Nature
       and the Poor?   118*
    *Does Ecuador Have the Fiscal Capacity for Sustainable
       Development?   119*

11. Resolving the Policy Crisis   *124*
    Nondiscriminatory Macroeconomic and Sectoral Policies   125
    Efficient Markets for Natural Resources and Other Inputs   126
    *Formation of Human Capital and Strengthening the Rural
       Economy's Scientific Base   130*
    The Challenge of Reform   131

Abbreviations   *133*

References   *137*

Index   *145*

I

THE CAUSES OF RENEWABLE RESOURCE DEGRADATION

1

Environmental Crisis in the Latin American Countryside

Latin America is richly endowed with natural resources. Major deposits of oil and bauxite, copper, and other minerals are scattered from the Rio Grande to the Tierra del Fuego. At Carajás, in northern Brazil, there is enough iron ore to supply the entire world for centuries. Gold and other precious metals are mined in a number of places.

The region's renewable resources are similarly abundant. Many highland valleys in Central America and the Andes are blessed with volcanic soils and ample precipitation. Nowhere on the face of the earth are conditions better for agriculture than in the Argentine Pampas, a region that combines a Mediterranean climate with the soils of the midwestern United States. Drawing on rich natural ecosystems, many countries produce and export large quantities of timber and fish.

Where commercial deposits of petroleum or metallic ores exist, extractive income can be invested in alternative forms of wealth: infrastructure, human capital, and so forth. To be sure, this used to be a rare event in Latin America. During the colonial era, gold and silver shipments from the New World fueled inflation in Spain and gilded more than a few European churches and palaces. Long after independence had been achieved, taxes paid by mining and oil firms remained nominal. By contrast, recent decades' exploitation of mineral riches has been relatively enlightened. Of course, resource rents have been dissipated in bloated governmental bureaucracies and subsidized parastatal corporations. One cannot deny, however, that extractive earnings have also given many countries the means to complete public works projects and to attack illiteracy and disease.

Compared with the transformation of mineral deposits into alternative forms of wealth, degradation of renewable natural resources is far more disturbing. At a minimum, it fails to qualify as sustainable development. How, for instance, can the conversion of biologically diverse rainforests into low-quality farmland possibly be reconciled with the Brundtland Commission's call for development that "meets the needs of the present without compromising the ability of future generations to meet their own needs" (WCED, 1987, p. 8)? By the same token, one cannot be complacent about accelerated soil erosion, which reduces the capacity of many countries to meet expanding demands for agricultural commodities.

To explain damage to the environment, one can think in terms of market failure (Baumol and Oates, 1988). Certainly, externalities are pervasive in Latin America. Owners of vehicles and factories in Mexico City and São Paulo, for example, pay little heed to their individual contributions to severe air quality problems. At an international level, nations with extensive tropical forests do not have strong incentives to consider how humankind as a whole is affected by agricultural land clearing.

Another view of the problem is Malthusian. Notwithstanding recent declines in birth rates, Latin America's population continues to grow rapidly. Surely, one might say, this is why the region's soils, water, and biological resources are under threat.

However, neither the market failure model nor the simple Malthusian perspective fully explains the environmental crisis in rural Latin America. Consider the problem of soil loss. Although erosion's external impacts can be considerable, farmers typically suffer most from erosion. Furthermore, some conservation measures are fairly inexpensive. Why, then, are those measures seldom adopted? Also, it cannot be denied that population growth causes resource scarcity to increase. But how can purely demographic explanations of environmental problems in Latin America be reconciled with the fact that there are many parts of the world where population density is higher and yet renewable natural resources are managed better?

Recognizing paradoxes like these, one is obliged to search for other explanations of depletive human interaction with the natural environment. We argue that waste and misallocation have to do primarily with the policies facing those who use and manage renewable resources.

The importance of public policy is widely recognized. For example, Blaikie (1985) contends that soil erosion in Africa, Asia, and Latin America is a consequence of development strategies that marginalize the rural poor, hence driving them onto fragile lands. We have no major quarrels with this analysis. In the pages that follow, however, a more inclusive thesis is offered, which is that renewable resource depletion is symptomatic of a policy regime that hinders economic activity in rural areas.

Outline of the Book

Careful documentation of this thesis requires that the scope of our study not be too broad. One option would be to show how similar policies underpin a single environmental ill (tropical deforestation, for instance) in a number of countries. We use a different approach. Various problems arising throughout Latin America are examined in one nation—Ecuador, which is wedged between Peru and Colombia in northwestern South America (Figure 1-1). Each of those problems is shown to result from policies that are applied throughout the region.

The study begins with a survey of forces influencing natural resource scarcity in the Latin American countryside, generally, and rural Ecuador, specifically. As is reported in Chapter 2, population growth will continue

Figure 1-1 Ecuador and neighboring countries.

for many years to come. In addition, the economy is beginning to rebound after a decade of stagnation. Consequently, consumer demand for crops, livestock, and other commodities is expected to rise. At the same time, commodity exports are increasing as discredited policies that protected urban industry and consumers are abandoned. As domestic and international opportunities open up, derived demand for land, water, and other environmental inputs to the rural economy is increasing.

As is stressed in the third chapter, the policy response to mounting resource scarcity has been inadequate. Since public sector claims on natural resources are excessive and governmental interference with private land rights is widespread, conservation incentives are weak. Macroeconomic distortions and public sector regulation of market forces tend to suppress the rural economy. As a result, soils, water, flora, and fauna, which are among the economy's principal assets, are managed poorly. In addition, spending on education, research, and extension (i.e., technology transfer) is generally inadequate. This strengthens the link between resource depletion and economic activity in rural areas.

Our thesis of inadequate policy response to growing demands on the natural environment is elaborated in six case studies.

The first study, presented in Chapter 4, addresses tropical deforestation, which arouses substantial debate and concern around the world. Along with examining the extent and environmental consequences of agricultural land clearing in Ecuador and elsewhere in Latin America, we stress that excessive conversion of primary forests into cropland and pasture is a clear illustration of how public policy has rewarded the "mining" of renewable natural resources. Deforestation is accelerated by the underpricing of trees and the land where they are rooted. Above all, encroachment on natural ecosystems is farmers' and ranchers' main response to growing demands for crops and livestock because productivity in the agricultural economy is extraordinarily low.

Farmland degradation is examined in the fifth chapter. Disequilibrium in rural input markets, resulting from the government's attenuation of private property rights and its repression of financial intermediation, is shown to weigh heavily on farmers, especially the owners of small hillside parcels. As a result, they are discouraged from adopting soil conservation measures and other land improvements.

The consequences of distorted prices for natural resource commodities are highlighted in Chapter 6. Since water and hydroelectricity have been sold at a small fraction of their cost, resources have been wasted on a grand scale. Ever larger and more inefficient dams, canal systems, and turbine complexes have been built. In addition, personnel and funding for pollution control and improved watershed management have been woefully inadequate.

Another issue arising in tropical rainforests is addressed in the seventh chapter. The environmental impacts of petroleum extraction in the

Ecuadorian Amazon are described, as is the international controversy provoked as new fields are brought on line. The economics of improved pollution control are also examined. Our analysis suggests that tax revenues might have to be reduced if the application of tighter environmental regulations is not to choke off private investment.

Shrimp mariculture is booming in many parts of Latin America and Asia. As we point out in Chapter 8, industry expansion in places like Ecuador has often been at the expense of the environment. Property arrangements encourage people to construct shrimp ponds in mangrove swamps and other coastal habitats. Continued growth in output as well as the prospects for natural resource conservation both depend on the adoption of improved technology, which reduces the industry's dependence on ecosystem mining.

The trade-offs to be struck between tourism development and environmental protection in unique natural habitats are diagnosed in the final case study. The singular and fragile Galápagos Islands are shown in the ninth chapter to be under threat largely because tourism has contributed much less than it could and should to ecosystem maintenance and recuperation.

Our analysis of the environmental crisis in the Ecuadorian countryside yields recommendations for a conservation strategy. Those recommendations are taken up in the final part of the book.

As we point out in Chapter 10, several myths about the causes of renewable resource degradation in places like Ecuador are widely circulated, both in poor countries themselves and in more affluent parts of the world. Those taking a pessimistic view make two arguments. First, population growth is out of control. Second, nothing can be done to ease the environmental impacts of mounting demographic pressure or economic expansion.

We reject both arguments. Evidence presented in the second chapter shows that rates of population increase are declining. In addition, environmental constraints need not prevent future generations from enjoying higher standards of living. Of course, environmental wealth could be dissipated as Ecuadorians struggle to feed, clothe, and house themselves. To avoid this outcome, natural resources need to be matched with more human-made wealth. True development, as opposed to living off the proceeds of environmental depletion, requires that goods and services be produced by combining natural resources and "nonenvironmental assets" (above all, human capital).

The policy regime needed to encourage a transition from ecosystem mining to genuine economic development in rural areas is outlined in the eleventh chapter. Secure property rights have to be vested in those who use and manage natural resources. Also, macroeconomic and sectoral policies that hamper rural development must continue to be wrung out. Furthermore, investment in human capital and the rural economy's scientific base is essential. All of these reforms, it must be emphasized, are well within the reach of Ecuador and its neighbors.

The Study's Geographic Focus

The topics addressed in this book are of transcendent importance, not just to the poor countries of the Western Hemisphere, but to humankind as a whole. In addition, how to balance economic progress and environmental protection is the subject of a large and growing literature. Our study contrasts with many contributions to that literature, which tend to take a "broad-brush" approach. Particularly in the six case studies, we offer specific insights on how to exploit complementarities between development and conservation.

There are two primary reasons why those insights, drawn from an in-depth study of a single country, are widely applicable. First, Ecuador is geographically heterogeneous, containing within its boundaries just about every major habitat in Latin America. Second, the natural resource issues it faces are entirely representative of those facing the rest of the region. In just one small nation, one can come to grips with the hemisphere's major environmental problems.

With 26.27 million hectares of territory on the South American mainland and 0.80 million hectares in the Galápagos Islands, Ecuador is about the same size as the United Kingdom or the state of Colorado. Yet it is an astonishingly diverse country, a place where major South American environments converge.

The desert that starts nearly 3,000 kilometers down the Pacific coast, just north of Valparaiso, Chile, runs up to the peninsula west of the port city of Guayaquil. The tip of the same peninsula, however, is only 150 kilometers away from the humid tropics. Rainforests used to stretch unbroken from a little southeast of Guayaquil, along Colombia's Pacific coast, through Central America, and up to the southern shores of the Gulf of Mexico.

If anything, geographical variety becomes even more dramatic as one moves inland. No wider than 300 kilometers, the Ecuadorian Costa is bounded on the east by the precipitous western range of the Andes, which is dominated by Mount Chimborazo (6,310 meters above sea level). In the Sierra, as in the coastal plains, there is a confluence of Latin American environments. The northern highlands are like the Colombian and Venezuelan Andes and many parts of Central America, with mountain chains hemming in a series of bowl-shaped valleys. With good soils and water availability, those valleys are excellent places to grow temperate-zone crops. By contrast, conditions for crop production are much less favorable in the southern Sierra, where dry uplands covered by infertile soils are cut in a few places by narrow river gorges.

After the eastern range of the Andes has been crested, a rapid descent into the Amazonian rainforests of the Oriente begins. A frontier city, Nueva Loja is only 400 kilometers from the Pacific Ocean and 300 meters higher than the surface of the Atlantic Ocean, which is more than 3,000 kilometers due east. Since precipitation and temperature are uniformly high all year, plant life is luxuriant.

Rarely is Ecuador in the international spotlight. Thus far, it has avoided serious political violence. Although the production of coca and its derivatives always threatens to increase, the country's drug problems are dwarfed by those encountered elsewhere in the Andes. Its hesitant steps toward liberalizing foreign trade and investment and controlling inflation have been puny compared to the reforms implemented in Chile and Mexico, for example.

When foreigners hear about the country, then, it is usually because some environmental problem afflicting all of Latin America has developed into an emergency in Ecuador. Its rainforests, which are home to countless endemic species, are threatened by accelerated agricultural land clearing and oil industry pollution. Dry forests in the southern and western parts of the country, which are also of enormous interest to biologists, are being lost because of overgrazing by livestock and excessive fuelwood collection. Fertile soils are sliding down hillsides and mangrove swamps are being bulldozed into shrimp ponds. Water is being mismanaged on a grand scale. The future of the Galápagos Islands, which are a precious heritage for all humankind, is uncertain.

Each of these problems is worthy of study in its own right. Furthermore, neither the causes nor the consequences of renewable resource deterioration in Ecuador are unique, as we emphasize in the pages that follow. Our findings, then, are of interest far beyond the country's borders.

2

Causes of Increasing Resource Scarcity

Pressure on the natural environment is intense in the Latin American countryside. Our analysis of why this is so begins with an examination of the causes of mounting renewable resource scarcity.

Throughout the developing world, demographic growth has been dramatic. Annual rates of population increase have approached or exceeded 3.0 percent in many African, Asian, and Latin American nations. Growth is tapering off as women choose to have fewer children. But median age remains low, which means that populations will not stabilize until well after the turn of the century.

The geographic incidence of demographic change is far from uniform. Human fertility is generally higher in rural areas. However, people have also been leaving the countryside in droves. In terms of the numbers involved, the recent expansion of third world cities reflects the largest migration in human experience. Agricultural frontiers are being colonized as well. Fewer people are converting forests into farms, although the impacts on renewable resources are dramatic.

Whether it occurs in urban or rural areas, population growth enhances derived demand for inputs used to produce food and other commodities. Human-made substitutes for environmental inputs are in short supply in most developing countries. Consequently, any addition in the number of consumers tends to translate into accelerated exploitation of soils, forests, and fisheries.

At various times, other forces have contributed to rising commodity demand and, hence, growing resource scarcity. Most Africans, Asians, and Latin Americans are poor, and so have high income elasticities of demand for food. Whenever per capita gross domestic product (GDP) goes up, then, consumption of agricultural products increases as well.

For several years after the onset of the debt crisis in 1982, income growth was sluggish at best in many developing nations. However, policies that previously discouraged production of commodities for domestic and international markets have been reformed. Where this has happened, the scarcity value of renewable resources has continued to rise.

As is reported in this chapter, recent decades' changes in derived demand for renewable resources in rural Ecuador have been fairly representative of what has taken place elsewhere in Latin America. After peaking at more than 3.0 percent a year during the late 1960s and early 1970s, population growth has declined steadily, to around 2.0 percent. Tens of thousands have moved to agricultural frontiers and Ecuador's cities have burgeoned. Per capita GDP has stagnated since the early 1980s. But the rural economy has benefited from the dismantling of policies that used to discourage crop, livestock, timber, and fish production.

Before turning to a review of demographic forces and market conditions, we must emphasize that the trends and events discussed in this chapter do *not* necessarily induce environmental depletion. As is made clear in the next chapter, renewable resources have been wasted and mismanaged in places like rural Ecuador primarily because national policies have discouraged a sound response to mounting scarcity.

Population Growth

Studies of environmental problems in Latin America frequently begin with a summary of demographic trends. This is certainly an appropriate point of departure for an analysis of factors influencing renewable resource scarcity in Ecuador.

For over half a century, the country's population has increased at a rapid pace. During the early 1950s, annual growth rates exceeded 2.8 percent. They peaked at 3.3 percent between the 1962 and 1974 censuses. From the early 1970s through the early 1980s, rates of increase were around 2.9 percent per annum (CONADE/INEC/CELADE, 1993, p. 21).

More recently, population growth has slackened. At the time of the 1990 census, Ecuador had 9.65 million people; since 1982, when the previous census was carried out, annual rates of increase have averaged less than 2.3 percent (CONADE/INEC/CELADE, 1993, p. 21). If they have not already declined to 2.0 percent, they will do so soon.

Before 1980, declining mortality explained the larger part of demographic change in Ecuador. Life expectancy rose by a third between the early 1950s and the early 1980s. Historically, birth rates have decreased more slowly than death rates. Between the early 1950s and the early 1970s, total fertility fell by less than 15 percent. During the next ten years, an additional 10 percent decline was registered (CONADE/INEC/CELADE, 1993, pp. 3–6). If anything, Ecuadorian birth rates have been above Latin American norms (Table 2-1). However, the recent deceleration in population growth strongly suggests that women are having fewer children. There are three underlying reasons why this is occurring.

First, Ecuador's population is more urbanized, as is indicated in the next section of this chapter. More concerned about living expenses and educating their offspring than adding another youngster to the family labor pool, city dwellers tend to have fewer children than their rural counterparts. In the late 1970s, for example, birth rates in Guayaquil and Quito were 65 and 59 percent, respectively, of the national average (INEC, 1984, p. 31). More evidence on the linkage between urbanization and declining fertility has been obtained in a survey conducted in 1989 by the Center for the Study of Population and Responsible Parenthood (CEPAR). Whereas the typical married woman in the countryside has had three children, her urban counterpart has had only two (CEPAR, 1990, p. 27).

For many families and individuals, a second reason for falling birth rates has been improved standards of living. Real per capita GDP has risen since 1965, as is documented later in this chapter. In Ecuador, as in other countries, middle-class and lower-middle-class households are much less inclined than poor ones to have many children. For example, CEPAR (1990, p. 38) found that poor families had nearly 50 percent more offspring than did middle-class families.

Table 2-1 Human Fertility and the Use of Contraceptives, Selected Latin American Countries

	Total fertility in 1989	Married women of childbearing age using contraceptives in 1987 (%)
Brazil	3.3	65
Colombia	2.9	63
Ecuador	4.1	44[a]
Guatemala	5.5	23
Honduras	5.3	41
Mexico	3.4	53
Paraguay	4.7	45
Peru	3.9	46

Source: IBRD (1991b), pp. 256–257.
[a]In 1986.

The third cause of declining birth rates is rising education levels. Of all the factors inversely related to a female's decision to bear children, none is more important than the number of years she has spent in school. On average, Ecuadorian women with no formal instruction have more than five children. By contrast, the typical high school graduate has one or two (CEPAR, 1990, p. 27).

Females who are better educated, more prosperous, and live in cities have been able to act on their desire to have fewer children because contraceptives have become more widely available. As of 1986, some 44 percent of married Ecuadorian women of childbearing age were using birth control measures (Table 2-1). This was low relative to what was observed in other Latin American countries (Table 2-1). But in the last few years, family planning services have become more available in Ecuador, especially in urban areas. CEPAR (1990, p. 51) estimates that 53 percent of the country's married women of childbearing age were using contraceptives in 1989. That portion has continued to increase.

Before returns to the 1990 census were published (INEC, 1991), there was universal agreement that Ecuador's population was growing by nearly 2.7 percent a year and had already passed 10 million. The news that the latter milestone had not yet been reached was greeted with considerable relief. Complacency about demographic pressure, though, is not in order. As of 1990, median age was just twenty years (INEC, 1991, p. 37), which means that the number of females of childbearing age is still increasing. If recent progress on controlling fertility is not sustained, Ecuador, which is the most densely populated country in South America (more than 37 residents per square kilometer), will get much more crowded during the next few decades.

Internal Migration

Although the number of Ecuadorians has increased rapidly in recent decades, rural population in most provinces (Figure 2-1) has remained fairly stable. Indeed, Bolívar, Imbabura, and Loja, in the highlands, actually lost rural population between the 1974 and 1990 censuses. So did two Costa provinces: El Oro and Manabí.

To understand why the rural population of the western two thirds of Ecuador was growing much more slowly than was the general population, one must keep in mind migration to the United States, which is mostly undocumented. In addition, demographic trends in the Oriente and in cities and towns have to be taken into account.

Consistent with what can be observed from Bolivia to Colombia, migration has taken place from the Andes to adjacent Amazonian lowlands. Demographic change has been especially pronounced in Napo and Sucumbios, where oil is produced. Whereas population growth in Ecuador's five Oriente provinces averaged 4.7 percent a year between

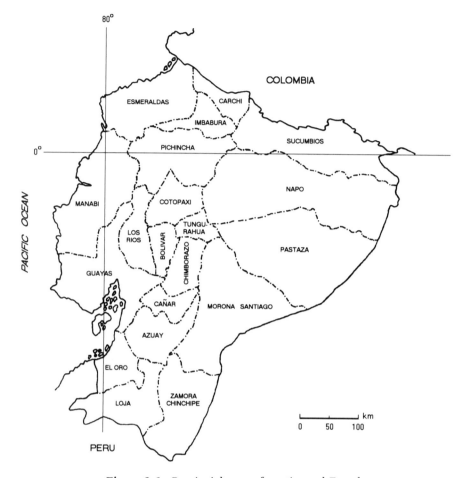

Figure 2-1 Provincial map of continental Ecuador.

1974 and 1990, annual rates of increase in the two northeastern provinces (which were split in 1989) have exceeded 6.0 percent since 1974 (INEC, 1991, p. 13).

To be sure, population growth in Amazonia has been rapid and undoubtedly has affected the natural environment. However, no demographic trend in Ecuador has been as impressive, in either absolute or relative terms, as relocation from the countryside to cities and towns. In 1974, the country's urban population was 2.70 million, equal to 40 percent of the total. That population grew by 4.2 percent per annum during the ensuing sixteen years, reaching 5.30 million, or 55 percent of the total (INEC, 1991, p. 35). Guayaquil and Quito have been expanding rapidly, as have several medium-sized cities.

Ecuador's transformation from a rural to a semiurban country has been pronounced even by Latin American standards (Table 2-2). Keeping this transformation in mind, one understands that population growth exerts pressure on Ecuador's natural resources in a largely indirect way—by enhancing derived demand for environmental inputs needed to produce agricultural and other commodities for ever increasing numbers of urban dwellers.

Income Growth

While critical, demographic growth is not the only reason why demand for agricultural commodities and other goods and services produced in the Ecuadorian countryside has risen in recent decades. At times, demand trends have also been influenced by rising incomes.

Income growth has been far from steady. All told, per capita GDP increased by nearly 3.0 percent a year during the past quarter century. But growth was largely confined to the period between the late 1960s and early 1980s, when the development strategy based on import substitution and industrialization was yielding large dividends. As Ecuadorian manufacturers expanded to meet demand in protected domestic markets, employment and income increased. Per capita GDP rose very rapidly during the petroleum boom, which began immediately before the run-up in prices triggered by the Arab oil embargo of 1973–1974. Windfall gains captured during the boom allowed for heavy subsidization of industry as well as major increases in public sector employment (Marshall-Silva, 1988), both of which tended to benefit urban areas and especially the middle classes in cities and towns. The economy expanded

Table 2-2 Urbanization in Selected Latin American Countries, 1980 to 1988

	Total population in 1989 (in millions)	Annual growth, 1980–89 (%)	Urban population in 1989 (in millions)	Annual growth, 1980–89 (%)
Brazil	147	2.2	109	3.5
Colombia	32	2.0	22	3.0
Ecuador	10	2.7	6	4.5
Guatemala	9	2.9	4	3.4
Honduras	5	3.5	2	5.5
Mexico	85	2.1	61	3.0
Paraguay	4	3.2	2	4.6
Peru	21	2.3	15	3.1

Source: IBRD (1991b), pp. 254–255, 264–265.

by nearly 7.9 percent a year from 1965 through 1981. With the population increasing at a little less than 2.9 percent a year, annual growth in per capita GDP approached 5.0 percent.

Conditions changed once petroleum prices fell and the debt crisis began, thereby shrinking the government's financial resources. In 1982, stagnation set in, as it did in the rest of Latin America, and expansion of the Ecuadorian economy has, at best, barely kept pace with population growth since then. Annual GDP growth averaged 2.1 percent from 1982 through 1989, while the population was increasing by more than 2.2 percent a year.

Increasing Domestic Demand for Agricultural Commodities

Changes in population and income since the middle 1960s having been surveyed, recent trends in domestic demand for agricultural commodities can now be examined. This analysis is done both for the time when economic growth was driven by public sector expansion and import substitution and industrialization and also for the subsequent period of austerity.

Simple estimates of change in demand for agricultural commodities have been obtained by adding population growth to the product of change in per capita GDP (a measure of average income) and the income elasticity of demand for food. Twenty-five years ago, per capita GDP was low and income elasticity was correspondingly high. Tschirley and Riley (1990, p. 196) suggest that income elasticity for agricultural commodities as a whole was between 0.40 and 0.60 during the 1970s. (That is, 10 percent growth in earnings could have been expected to lead to a 4 to 6 percent increase in demand for the typical food item.) As per capita GDP has risen, income elasticity has probably fallen.

The information and assumptions used in our analysis of demand trends for the agricultural sector are summarized in Table 2-3. As is indicated there, 2.87 percent population growth combined with annual increases in per capita GDP averaging 4.99 percent and an income elasticity of 0.55 to induce demand growth of 5.62 percent per annum for the period 1965 through 1981. Once austerity set in, though, contractions in average income averaging 0.10 percent a year held down demand growth, which was still positive because the country's population continued to increase.

One should keep in mind that estimates of demand growth presented in Table 2-3 represent trends for the agricultural sector as a whole. Normally, changes in consumption induced by rising incomes vary considerably across subsectors and individual commodities. Of particular importance in developing countries, including Ecuador, is that income elasticities for livestock products tend to be relatively high.

Table 2-3 Factors Contributing to Increased Domestic Demand for Food, 1965 to 1989

	Population growth[a] (%)	Income growth[a] (%)	Income elasticity[b]	Demand growth[c] (%)
1965–81	2.87	4.99	0.55	5.62
1982–89	2.24	−0.10	0.45	2.20
1965–89	2.65	2.95	0.50	4.13

[a] Authors' analysis of reports issued by the Central Bank of Ecuador (BCE) and of census returns (CONADE/INEC/CELADE, 1993, and INEC, 1991).
[b] From Tschirley and Riley (1990), p. 196.
[c] Population growth + (income growth × income elasticity).

All else remaining the same, increasing food consumption can be expected to enhance derived demand for natural resource inputs to agricultural production. This has certainly been the case in Ecuador, where modern, human-made inputs, which can substitute for natural resources, are extremely scarce (Chapter 3). However, as is documented in the next section of this chapter, macroeconomic and sectoral policies have played a major role in conditioning the response to commodity demand trends. During the period of rapid economic expansion, policies favoring industry and urban areas led to relatively slow growth in the agricultural sector. This, in turn, tended to limit pressure on renewable natural resources. By contrast, growth in demand for agricultural products, and therefore pressure on resources, picked up during the austerity period in spite of stagnating incomes and reduced demographic pressure, largely because the same restrictive policies were gradually being abandoned.

Macroeconomic and Sectoral Policies and Performance of the Agricultural Economy

By the late 1960s, Ecuador was committed to an economic development strategy predicated on import substitution and industrialization (Schodt, 1987, pp. 89–94, 105–112). Under that strategy, which was pursued at one time or another in every Latin American nation (Krueger, Schiff, and Valdés, 1988), manufacturers and urban consumers were subsidized, through direct controls on the prices of food and raw materials, through currency overvaluation, and through trade restrictions. The resulting bias against agriculture was rarely, if ever, overcome by policies designed to help the sector (Scobie, Jardine, and Greene, 1990).

The bias was strong in the case of Ecuador's three traditional export crops: bananas, cocoa, and coffee. Duties on foreign shipments of those commodities were gradually reduced, along with other taxes, during the 1970s, when oil exports became the primary source of public sector revenues (Marshall-Silva, 1988). However, overvaluation of the Ecuadorian sucre, which was growing progressively worse during the same period (Scobie, Jardine, and Greene, 1990), drove down domestic prices, which discouraged production. As is reported in Table 2-4, Ecuadorian output of bananas, cocoa, and coffee grew by just 1.2 percent a year between 1965 and 1981.

Producers of other crops fared little better than those who cultivated the three major export commodities. Along with currency distortions, subsidization of wheat imports discouraged domestic output. Between 1965 and 1981, annual growth amounted to only 2 percent (Table 2-4), which is far below increases in demand observed during the same period (Table 2-3).

In the livestock subsector, some farmers benefited from limited protection given to the dairy products industry. But performance of the subsector as a whole compared poorly with consumption trends. Between 1965 and 1981, output grew by nearly 4.6 percent a year, which is less than the estimate of sector-wide demand growth, 5.6 percent, for the same period (Table 2-3). That estimate is, in turn, lower than actual consumption increases because income elasticity of demand for livestock products is generally higher than demand elasticities for other foodstuffs. Ecuador was satisfying its increasing appetite for meat, milk and cheese, poultry, and eggs between the late 1960s and early 1980s largely by purchasing more of those commodities from foreigners.

The situation for other parts of the rural economy was quite different between the late 1960s and early 1980s. The fisheries sector performed very well primarily due to the rapid expansion of shrimp mariculture. Raising shrimp in ponds was profitable in spite of adverse macroeconomic policies both because output was being sold in an expanding global market and because maricultural enterprises could exploit highly productive coastal ecosystems at virtually no cost to themselves (Chapter 8). The forestry sector is defined to include the wood products industry,

Table 2-4 Annual Percentage Growth Rates in the Rural Economy, 1965 to 1989

	Bananas, cocoa, and coffee	*Other crops*	*Livestock*	*Fishing*	*Forestry*
1965–81	1.21	2.05	4.56	11.73	9.61
1982–89	2.53	7.69	1.85	15.81	2.36
1984–89	3.85	8.26	1.40	16.45	3.12

Source: BCE reports.

which has benefited from measures (e.g., a ban on log exports) designed to keep the prices of its raw materials low.

Once the impacts of declining oil prices and the debt crisis began to be felt, fundamental changes in macroeconomic and sectoral policies became obligatory. Currency devaluations began in the early 1980s, during the administration of President Osvaldo Hurtado. Opening of the Ecuadorian economy accelerated while León Febres Cordero was president, from 1984 to 1988. Tariffs were reduced and flexible, market-based exchange rates were introduced. Currency distortions and tariffs continued to be wrung out during the government of Rodrigo Borja, whose four-year term ended in August 1992. A social democrat, President Borja often seemed to be an unwilling participant in this process. As was the case in previous administrations, recent advances toward an outward-oriented development strategy have been uneven (Scobie, Jardine, and Greene, 1990).

Ecuadorian agriculture has responded to the improved policy environment. Factoring out the major damage caused by El Niño storms in 1983 (see third row of Table 2-4), one observes that output of the three traditional commodity exports grew. Even more dramatic were contemporary increases in the production of other crops. Once currency distortions (which made imports cheap) had been reduced and subsidies for wheat imports had been removed, Ecuador's farmers began to reclaim domestic markets and take advantage of opportunities to sell abroad. Production of crops other than bananas, cocoa, and coffee increased by 8.3 percent a year from 1984 through 1989 (Table 2-4). Agricultural exports grew by 11.4 percent a year between 1983 and 1988. This rate of annual increase was exceeded in only two other Latin American countries: Chile, with 17.5 percent a year, and Mexico, with 14.0 percent a year (Southgate, 1991).

Falling incomes during the 1980s dampened demand for livestock products. Also, controls on the prices of selected goods (e.g., milk) continued to be enforced. Consequently, output grew slowly (Table 2-4).

With shrimp mariculture continuing to grow, the fisheries sector's upward trajectory accelerated in the 1980s. By contrast, the forestry sector had benefited from subsidies inherent in import substitution and industrialization policies and also lost important markets in the Andean region early in the decade. As a result, its performance during the 1980s compares poorly with what was observed from the middle 1960s to the early 1980s (Table 2-4).

The process of policy reform that began in the early 1980s has had a positive impact on Ecuador's rural economy. [For a review of production trends for specific agricultural commodities during the last quarter century, see Whitaker and Alzamora (1990b).] As land use trends documented in the fourth chapter of this book indicate, production increases have, in turn, enhanced derived demand for environmental inputs in the countryside.

The Challenge of Increasing Scarcity

As is indicated in this chapter, the forces contributing to increased natural resource scarcity in rural Ecuador have been strong. Mounting demographic pressure has been the most important factor causing demand for food and other commodities produced in the countryside to increase. At times, demand has been stimulated by income growth. In recent years, macroeconomic policies have become more outward-oriented, which has caused farmers and other rural dwellers to produce more output for domestic as well as international markets. All of these trends have translated into increased derived demand for land, water, and other natural resource inputs to the rural economy.

Mounting resource scarcity does not have to result in accelerated environmental degradation. If secure property rights are vested in resource users and if market prices are in line with scarcity values, then conservation measures will be adopted as derived demand for natural inputs increases. In addition, resource scarcity can be alleviated through investment in human capital and the rural economy's scientific base. Both types of investment allow for crops, livestock, timber, and fish to be produced with more human-made factors and, thus, fewer environmental inputs.

The next chapter's analysis makes clear that, unfortunately, Ecuador has been slow to put in place the set of policies needed for successful accommodation of growing demands for renewable natural resources.

3

Policy Crisis and Environmental Degradation

Does mounting derived demand for land, water, and other resource inputs to the rural economy imply that environmental degradation must accelerate? Certainly not. Indeed, the opposite result is entirely conceivable. That is, people might use and manage resources more, not less, carefully as scarcity values rise. Unfortunately, this response is unlikely under the policy regime applied in many Latin American nations, including Ecuador.

By suppressing economic activity in the countryside, distorted macroeconomic and sectoral policies have probably caused some renewable resources to sit idle. But at the same time, those policies have weakened conservation incentives and have discouraged the use of human-made substitutes for environmental inputs.

Poor performance of rural land and financial markets is also the result of public policy and has adverse consequences for renewable resources. Real estate markets fail to communicate scarcity values because public sector claims on resources are excessive and because the government interferes too much with private land rights. Interest rate controls are largely a thing of the past. However, chronic inflation inhibits financial intermediation, which resource users need to invest in conservation measures and also to capture the long-term benefits of sustainable development.

Renewable resource depletion is all but inevitable wherever the rural economy's scientific base is weak. Research and extension in support of agriculture, forestry, and other sectors are severely underfunded in most Latin American countries. As a result, opportunities to substitute away from natural resource inputs are foreclosed.

The agricultural sector's inadequate response to mounting resource scarcity is the primary focus of this chapter. Discriminatory macroeconomic and sectoral policies, distorted financial and real estate markets, and underinvestment in the scientific base for crop and livestock production are all shown to result in resource depletion.

We also discuss why people who eke out a meager living by mining soils and natural ecosystems are prevented from switching to alternative forms of employment. Labor market restrictions are part of the problem. But more than anything else, inadequate spending on education keeps the rural poor in their place, with dire consequences for the environment.

This chapter's general analysis lays the groundwork for an examination of tropical deforestation, soil erosion, and other specific natural resource issues in the next part of the book. It also sets the stage for a discussion of policy reform, which should be the centerpiece of conservation strategies for places like Ecuador.

Discriminatory Macroeconomic and Sectoral Policies

As is emphasized in the preceding chapter, macroeconomic distortions and related interference with market forces have long been major features of public policy throughout Latin America. Currencies have been overvalued and tariff and nontariff barriers have been erected to protect domestic industry from foreign competition. Bowing to urban and industrial interests, governments have kept prices for agricultural and natural resource commodities low, through direct price controls as well as discriminatory macroeconomic and trade policies (IICA, 1988; Valdés, 1986). Although many of these policies have been changed in Ecuador, some remain in place.

The environmental impacts of holding down the prices of crops, livestock, timber, and other commodities are mixed. Diminished economic activity in the countryside lowers renewable resource use. For example, cropland did not increase very much during the 1970s (Chapters 4 and 5) because adverse macroeconomic and sectoral policies discouraged farmers from responding to rapidly expanding demand for foodstuffs (Chapter 2).

At the same time, though, incentives to conserve resources are weak when policies are tilted against the rural economy. It is more difficult to justify expenditures on erosion control and water management, for example. By the same token, there is a temptation to mine the environment, which is a fixed asset as far as resource users are concerned. As we show in Chapter 5, farmland degradation is a case in point.

Just as it works against resource conservation, a policy-induced decline in the agricultural sector discourages the adoption of improved technology for crop and livestock production. For example, commercial fertilizers are not used by many of Ecuador's farmers. Instead, they find it in their own best interest to exploit the environment for cheaper substitutes.

Virtually all research on the consequences of distorted macroeconomic and sectoral policies for Latin American agriculture has focused on production impacts. To understand how those policies influence renewable resources, comprehensive assessment of the performance of rural input markets is needed. Anticipating the results of this assessment, we have no doubt that, while policies biased against agriculture can cause a smaller portion of the rural economy's environmental base to be employed, management of that portion often turns out to be highly depletive.

Factor Market Distortions

As Ecuador abandons its long and costly experiment with import substitution and industrialization, another set of policy-induced distortions becomes relatively more important. Governmental intervention in input markets is pronounced in the country, taking several forms and creating various environmental impacts. As is reported in Chapter 6, water and hydroelectricity subsidies cause resources to be wasted, misallocated, and polluted. In the next few pages, we address the environmental consequences of poor performance in rural land and financial markets.

Land Markets: Inappropriate Property Arrangements and High Transactions Costs

In any economy, real estate markets perform a critical function, which is to transmit scarcity values to resource users. Unfortunately, this function is suppressed in Ecuador, largely because there is a gross imbalance between what the government claims as its own and what it is actually able to manage.

Public sector holdings are extensive. Subsurface resources have always been under governmental control. Coastal wetlands are "national patrimonies." With passage of the 1972 Water Law, all water resources were nationalized. In addition, most of the country's tree-covered land is included in a park or nature reserve or some other part of the Forest Patrimony of the State (Patrimonio Forestal del Estado). Nowhere else in the world does the central government claim to "protect" a larger portion of the national territory (Table 3-1).

More often than not, public sector agencies are unable to control access to their properties, which means that trespassing is widespread.

Table 3-1 Ten Countries with the Highest Ratio between Protected Areas and National Territory

	Ratio (%)
Ecuador	38.4[a]
Austria	19.3
Bhutan	18.6
Botswana	17.7
Panama	17.3
Chile	16.0
Czechoslovakia	15.8
Norway	15.5
Tanzania	13.4
Costa Rica	12.0

Source: IBRD (1991b), pp. 268–269.
[a] Includes parks, reserves, and other parts of the Forest Patrimony, much of which is occupied by indigenous groups and agricultural colonists.

Individuals regard the resources they use as a free (i.e., valueless) good, neglecting the social costs of environmental degradation. This is the crux of what Hardin (1968) calls the "tragedy of the commons," which describes well the state of many nationalized resources in Latin America.

In many cases, the Ecuadorian government tacitly recognizes that its claims on natural wealth are excessive relative to its capacity to control access. It chooses, then, to divest itself of its properties. Unfortunately, the policy arrangements governing divestiture have been inconsistent with the wise use and management of natural resources.

For one thing, the public sector has tended to sell off its holdings too cheaply. Among examples of concessionary pricing are the nominal charges paid by loggers extracting timber and agricultural colonists settling in the Forest Patrimony (Chapter 4), the low annual fees paid by maricultural enterprises constructing shrimp ponds in coastal wetlands (Chapter 8), and the negligible taxes that have been levied on Galápagos cruise ships (Chapter 9).

In addition, private rights in denationalized resources have not been structured to encourage conservation. Indeed, destruction of natural vegetation was long a prerequisite for formal or informal tenure in private holdings carved out of national patrimonies, be they tropical forests or mangrove swamps. Land rights have been won the same way in other parts of Latin America. Agricultural colonists in the Brazilian Amazon, for example, have obtained titles in forested land simply by clearing it (Mahar, 1989).

Obliging settlers to destroy natural vegetation causes clearing to take place any time agricultural income can be captured by doing so. This is because colonists realize that any delay in demonstrating formal or informal use rights through land clearing leads to the risk that somebody else will jump the claim (Southgate, 1990a). Furthermore, the terms governing the transfer of national patrimonies influence the development of resources already in private hands. Since users of the latter resources know that close substitutes can be acquired at low prices, their incentives for conservation are weak.

Excessive nationalization is not the only aspect of the property rights crisis responsible for a large part of environmental degradation in the Ecuadorian countryside. Private land rights are attenuated in all parts of the country, which causes natural resources to be managed poorly.

The rural titling system is chaotic (Lambert et al., 1990). Instead of being defined by boundary lines that have been properly surveyed, the typical holding is merely described in terms of neighboring properties and their owners. Furthermore, most registries comprise nothing more than chronological accounts of recorded sales and inheritances and many transactions are never recorded. Under these circumstances, verification of title is all but impossible (J. Rosholt, personal communications, 1990), which means that the institutional framework needed for modern land markets to function does not exist.

As is the case throughout Latin America, property rights attenuation in Ecuador also has much to do with agrarian reform. To be exempted from the general ban on renting land, which was enacted in the early 1970s in the hopes of ending "precarious" forms of tenure, an owner must prove that he or she is physically incapable of farming (e.g., because of illness). Also, the Ecuadorian Institute for Agrarian Reform and Colonization (IERAC) has a legal mandate to review all real estate transfers, which involves a complicated process (Seligson, 1984). Because the agency's record-keeping system is extremely cumbersome (IERAC did not acquire its first computer until the late 1980s), adjudication can take years. In the meantime, the rights of resource users are insecure.

The most damaging consequences of agrarian reform relate to expropriation risks. According to Ecuadorian law, land not fulfilling its "social function" can be taken from its current owner. In practice, redistribution follows a finding by an agrarian reform official that the parcel in question is being used inefficiently (e.g., that yields are too far below the national average) or that resources are not being conserved. Almost always, such findings are highly subjective. Even a well-managed farm can be seized if a representative of IERAC determines that population pressure is "high" in the surrounding area.

The displaced owner does not receive the current market value for the real estate. Instead, he or she is given agrarian reform bonds that can be cashed in only after several years have elapsed. The redeemed value of Ecuadorian bonds, which pay around 5 percent nominal interest, is roughly equivalent to the nominal value of property assessments carried out ten years before the expropriation. Obviously, this compensation amounts to a tiny share of real estate values given the 50 percent annual inflation that Ecuador has experienced for several years.

Since losing one's land through an agrarian reform action is so costly, many owners sell their holdings at bargain prices when faced with a real or threatened invasion—that is, the physical occupation of land by a group of people, which can trigger governmental involvement. Small fortunes have been made organizing such invasions and subsequently dividing up holdings extorted from the previous owners.

Conceivably, expropriation risks discourage some individuals from allowing their land assets to depreciate. However, a much more important impact of land reform has been to discourage conservation of those same assets. That is, owners are unlikely to invest in, say, erosion control measures if they are afraid that they will be displaced entirely from their property through an invasion, the action of an agrarian reform agency, or both.

Financial Sector Repression

Like policies responsible for the inefficient performance of real estate markets, repression of financial markets has a long tradition in Ecuador. Interest rate ceilings were first introduced in the 1930s. But because inflation was usually kept under control, those ceilings rarely created serious credit shortages. The situation changed in the early 1970s. Flush with oil wealth, the government began dispensing large amounts of cheap credit, an action that fueled inflation. Through direct and indirect controls, interest rates were consistently pegged below inflation rates. Since 1982, oil revenues have stagnated and lending from foreign banks has evaporated. As a result, credit shortages have grown severe (Ramos and Robison, 1990).

The problems of the financial sector are not just a consequence of interest rate policy. Property rights attenuation also impedes efficient intermediation between savers and borrowers. Since lending institutions cannot use rural property registries to verify titles reliably and cheaply, they must instead inspect entirely on their own all characteristics of any holding being offered as collateral. Given the costs implied by this arrangement, most agriculturalists find it hard to get the credit needed to cover the short-term costs of soil conservation measures and other land improvements. Financial intermediation is also restricted because of the transactions costs and expropriation risks that are the principal legacy of agrarian reform legislation.

The rural elite is not greatly inconvenienced by this state of affairs. Among others, Gonzalez-Vega (1984) has shown that lenders prohibited from charging more than a certain interest rate tend to allocate credit to large farmers, whose property rights are relatively well documented. Among other things, this allows bankers to reduce the transactions costs associated with making loans. For example, only 10 percent of Ecuador's 700,000 farmers and ranchers receive agricultural loans from the state-owned National Development Bank (BNF) and half of all credit disbursed by that institution goes to just a tenth of its borrowers, or 1 percent of the country's farmers (Camacho and Flores, 1993).

Denied access to artificially scarce formal credit, most small farmers rely on informal financial markets, in which interest rates are considerably higher than what rates would be in efficient formal markets. Having to pay the higher rates, they find it difficult to justify investment in resource conservation measures and other land improvements. Furthermore, the lack of financial intermediation makes it hard to internalize the long-term benefits of such improvements, in the form of higher real estate values.

Failure to Invest in the Rural Economy's Scientific Base

Even if resource users had nonattenuated property rights and if governmental intervention in product and input markets were eliminated, conservation of Ecuador's renewable natural resources would not be guaranteed. As long as the rural economy remains poor in everything but land, water, flora, and fauna, those resources will continue to be exploited.

Consider, for example, the consequences of a weak scientific base for crop and livestock production. As Whitaker (1990, pp. 313–314) documents, real spending on agricultural research in Ecuador has declined 7.3 percent a year from 1975 through 1988, when it fell below 0.2 percent of agricultural GDP. This level of support is low even by the standards of neighboring countries, few of which follow the recommendation of multilateral donors to spend 1.0 to 2.0 percent of sectoral GDP on research. Since limited research is matched by a weak extension service, crop yields in Ecuador are well below the region's middling standards (Table 3-2). Under these circumstances, increasing demands for agricultural commodities must be met primarily by bringing additional land, which is usually of marginal quality (Chapter 4), into production.

The relationship between low agricultural productivity and the sector's geographic expansion is emphasized by Whitaker and Alzamora (1990b). They point out that about two thirds of the increased crop production occurring in Ecuador between the middle 1960s and the middle 1980s was accounted for by bringing more land into production; improved yields explained only the remaining third (Whitaker and Al-

Table 3-2 Crop Yields in Colombia, Ecuador, and Peru during the Late 1980s (kg/ha/yr)

	Colombia	Ecuador	Peru
Barley	1,908	889	1,170
Corn	1,360	1,369	1,962
Peanuts	1,442	893	1,892
Paddy rice	4,697	2,302	4,809
Soybeans	1,998	1,890	1,795
Wheat	1,740	972	1,278

Source: From Whitaker (1990), p. 304.

zamora, 1990b, pp. 136–143). In a statistical analysis, Southgate (1991) demonstrated that a strong inverse relationship between agricultural productivity and frontier expansion applies throughout Latin America. The linkage between encroachment on natural ecosystems and low spending on research and extension also holds for the Ecuadorian shrimp industry (Chapter 8).

To be sure, formation of nonenvironmental assets should reflect a rural economy's factor endowments. For example, development and dissemination of yield-enhancing technology for crop and livestock production does not usually occur where land and other natural resources are abundant (Hayami and Ruttan, 1985). Unfortunately, Ecuador is investing very little in agricultural research and extension, even though its renewable resources are being stretched very thin.

Labor Markets and the Environment

As in many other parts of Latin America, agricultural development in Ecuador requires that renewable resources be complemented by more human capital and a stronger scientific base for crop and livestock production.

It is important to point out, however, that agriculture is unlikely to absorb all the labor available in the countryside. For living standards to improve and for soil, water, and natural ecosystems to be used and managed sustainably, jobs must be created in other sectors of the economy. For many displaced farmers and agricultural laborers, the best choice will be to work at higher wages in cities and towns, as hundreds of thousands of Ecuadorians already have done (Chapter 2). In addition, the potential for nonagricultural employment in rural areas is great, especially if opportunities for agribusiness development are exploited.

Some policies impede the exit of labor from nonremunerative work on farms. Ecuador has a minimum wage law and, since 1938, has applied a labor code that gives substantial benefits to workers. To discharge somebody who has been on the job for ninety days or more, a firm has to offer generous severance pay. In addition, workers at an establishment employing more than thirty people have the right to organize a union that can, in turn, strike for any one of a number of reasons, including sympathy for another union's job action or opposition to the government's economic policy.

Of course, this regime reduces the demand for labor (Hachette and Franklin, 1990). When filling a position, an employer has to consider the financial consequences of firing a worker at some future date. There are also many small enterprises in the country that are reluctant to undertake what would otherwise be an efficient expansion (e.g., to capture economies of scale) because they do not want to deal with unions.

A more serious impediment to intersectoral labor movement is inadequate formation of human capital. Literacy rates, which are one indicator of a population's earning power, are much lower in the countryside than in cities and towns. As of the 1990 census, 24 percent of all rural people aged ten years or older were judged to be unable to read. For the same age cohort in urban areas, the estimated illiteracy rate was 6 percent (INEC, 1991, pp. 34, 67). If anything, these estimates are low.

During the past quarter century, Ecuador has made a large investment in primary and secondary education. But there is still much room for improvement, especially in rural areas. Whitaker (1990, p. 307) points out that half the children in the countryside drop out of school before reaching the sixth grade. One has to infer from this fact that people suited for only the most menial and poorly paid work will continue to comprise a large part of the rural population for some time to come.

If labor market rigidities are not removed and if education levels remain low in the countryside, many rural Ecuadorians will find that there is no alternative for them but to try to survive on a small holding. Since most such holdings are in areas prone to soil erosion or where natural ecosystems are under threat, the probable environmental consequences of this outcome are grim.

An Ideal Set of Policies for Environmental Degradation

Like the rest of Latin America, Ecuador is a considerably different place from what it was only a few decades ago. Well into the current century, a large part of the country's population still dedicated itself to tilling the rich soils of inter-Andean valleys. Most of the Costa remained forested. The Oriente's contacts with the outside world were generally tenuous.

Many of the policy arrangements described in this chapter are a vestige of those simpler times, when the best way for a small population to feed, clothe, and house itself was to extract natural resources. For example, the frontier tenurial regime induced farmers and ranchers to respond in an entirely appropriate way to increasing demands for agricultural commodities—by converting forests into cropland and pasture. Likewise, formation of human capital and development of a scientific base for the rural economy made little sense when that economy could draw on a rich endowment of natural wealth.

Rarely does policy reform anticipate changed conditions. To the contrary, reform usually occurs only after the need for it has become compelling, often because a crisis has arisen. So it is that, in Ecuador and in many other Latin American countries, agricultural land use conversion continues to be encouraged in the face of adverse natural conditions, both because a frontier property regime remains in place and because human-made inputs that substitute for land and other natural resources in the production of crops and livestock remain scarce.

Not all the policy arrangements contributing to excessive pressure on Ecuador's renewable natural resources pertain to a time when the country was lightly populated. Food price controls, interest rate ceilings, and related distortions were implemented or became more severe between the late 1960s and the early 1980s, when many believed that public officials could improve considerably on the market's allocation of goods and services. It seems clear that the misguided efforts of government have accelerated environmental degradation in the countryside by reducing the rural poor's access both to credit and to better jobs outside of agriculture.

Because of the policy arrangements outlined in this chapter, Ecuador's agricultural economy and also the country's rural poor continue to depend on renewable resource extraction. As is demonstrated in the case studies reported in the next part of the book, the costs associated with this dependence are very high.

II

CASE STUDIES

4

Tropical Deforestation

The fate of tree-covered areas close to the equator is important to all humankind. As forests are cleared to make way for new cropland and pasture, carbon dioxide and other greenhouse gases are released into the atmosphere. The earth grows warmer as a result, all else remaining the same (Detwiler and Hall, 1988). Biological diversity is also diminished since many of the world's plant and animal species live only in tropical forests (Wilson, 1988).

In addition to causing global warming and species extinction, ecosystem destruction in Africa, Asia, and Latin America carries a direct human cost. Some native forest dwellers have dealt with missionaries, traders, and mineral prospectors for more than a century. But contact between some groups and the rest of the world was minimal until very recently. For many people, recent decades have brought economic exploitation and the loss of cultural identity, not to mention exposure to outsiders' diseases.

In this chapter, the history and current rate of agricultural land clearing in Ecuador are reported and the benefits and costs of diminished tree cover are reviewed. We also offer a causal analysis of deforestation, the case being made that farmers' and ranchers' encroachment on natural habitats in the northwestern and eastern parts of the country is a classic example of renewable resource depletion brought about by bad public policy.

Historical Trends

Conversion of tropical moist forests into cropland and pasture began a long time ago in Ecuador. Before independence was achieved from Spain, farmers started to move inland from Guayaquil. By the 1870s, periodic yellow fever outbreaks had been contained and cocoa plantations were being established along waterways, which were the principal mode of transportation, as far as 75 kilometers north and east of the port city. By the turn of the century, at the height of Ecuador's cocoa boom, the extensive margin of coastal agriculture had been pushed another 75 kilometers north. The boom collapsed after World War I and deforestation was relatively slow during the 1920s and 1930s (Bromley, 1981).

Frontier expansion resumed after World War II and accelerated in the 1960s (Figure 4-1). Completion of the highway running through Santo Domingo marked the initiation of settlement in northwestern Ecuador. Dodson and Gentry (1991) report that untouched primary forests remain intact only in those few places west of the Andes that are totally inaccessible.

Until the 1960s, the Ecuadorian Amazon was a backwater. A few small towns had been established in valleys at the base of the mountains and there were religious missions and trading posts farther east. But indigenous groups comprised the majority of the population and their contact with the outside world was minimal.

All that changed in 1967, when oil was discovered in the northeastern part of the country. Just as an earlier generation of agricultural colonists followed navigable waterways inland from Guayaquil, tens of thousands of immigrants from the Costa and Sierra have settled along roads leading to oil fields.

Changes in land use in tropical Ecuador over twenty years are reported in Table 4-1. Between the early 1970s and the late 1980s, cropland west of the Andes increased by nearly a fifth, to more than 1.25 million hectares. Meanwhile, the area planted to annual and perennial crops in the Amazon rose by 350 percent, from 30,000 to 135,000 hectares.

As a rule, deforested land is dedicated to cattle production. Nearly 2 million hectares of new pasture have been created in the five provinces of western Ecuador (El Oro, Guayas, Los Ríos, Manabí, and Esmeraldas) since 1972. Another 0.50 million hectares were deforested to make way for livestock herds in the five Oriente provinces (Zamora Chinchipe, Morona Santiago, Pastaza, Napo, and Sucumbios). Table 4-1's data on pasture trends do not reflect agricultural colonization of the western lowlands of Pichincha, Cotopaxi, and other mainly Andean provinces, which was under way by the 1960s and continues to this day. All told, at

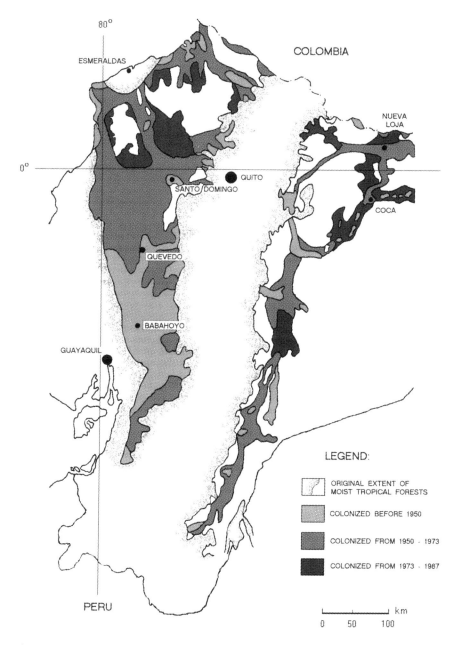

Figure 4-1 Tropical deforestation in Ecuador: a historical view. (Sources: Bromley, 1981, and SUFOREN reports)

Table 4-1 Agricultural Land Use Trends in Tropical Ecuador

Land use	1972–73 (ha)	1988–89 (ha)	Change (%)
Cropland	1,090,000	1,393,000	28
Costa	1,060,000	1,258,000	19
Oriente	30,000	135,000	350
Pasture	1,217,000	3,672,000	202
Costa	833,000	2,792,000	235
Oriente	384,000	880,000	129

Source: Reports of the Ministry of Agriculture and Livestock (MAG) and the National Agricultural Statistics System of the National Institute of Statistics and Censuses (SEAN-INEC).

least 3 million hectares of tropical forests have been converted into cattle pasture in the last two decades.

Current Magnitude

Around the developing world, deforestation rates are the subject of controversy. Ecuador is no exception. In particular, there is reason to believe that the most widely cited estimate of agricultural land clearing for the country—340,000 hectares a year (WRI, 1990, p. 42)—is excessive.

For one thing, the same source appears to have overestimated deforestation rates in other countries. For example, WRI (1990, p. 42) claims that 9.05 million hectares are being cleared each year in Brazil while the general consensus is that deforestation rates have fallen to 3 million hectares per annum or less (Pearce, 1991).

Another basis for challenging WRI's figure for Ecuador is Schmidt's (1990) estimate that land clearing throughout the Oriente has fallen to 60,000 hectares a year. In addition, there is the national government's survey of changes in forest cover between the middle 1960s and the middle 1980s. During that twenty-year period, average losses of tree cover in Ecuador's five eastern provinces were a little more than 100,000 hectares per annum, or a little more than 1 percent a year (SUFOREN, 1991, p. 42). This is generally consistent with Schmidt's (1990) findings since current deforestation rates are probably below the peak levels observed more than ten years ago.

When land clearing estimates for the Oriente and northwestern Ecuador are added to deforestation in the Guayas River Basin and elsewhere in the Costa, the estimate for the nation as a whole currently provided by officials of the newly created Ecuadorian Institute of Forestry, Natural Areas, and Wildlife (INEFAN)—140,000 hectares per annum—seems credible.

Net Returns

By no means does encroachment on natural ecosystems represent an unmitigated economic waste. Almost the entire area between the Appalachian Mountains and the Mississippi River, where a large share of the agricultural commodities traded in international markets is raised, was covered with trees and prairie grasses when farmers of European extraction first ventured into the region, two hundred years ago. Similarly, millions of hectares in southern Brazil that have long been used to produce coffee, soybeans, and other crops once were forested. In both cases, agricultural land clearing has proven to be beneficial.

The same can be said of a good deal of past deforestation in Ecuador, particularly in the Guayas River Basin. Seasonal flooding is a problem in some areas and soil erosion occurs in the upper watershed. In general, though, the region upstream from Guayaquil is an excellent site for crop and livestock production.

Unfortunately, most of the soils now traversed by Ecuador's agricultural frontiers do not lend themselves to farming or ranching. Furthermore, clearing results in major opportunity costs. As this section's analysis makes clear, the economic losses caused by current deforestation far exceed its gains.

Benefits of Agriculture's Geographic Expansion

The benefits associated with carving cropland and pasture out of natural ecosystems comprise farmers' and ranchers' income. That income, in turn, depends on how well their land is suited to agriculture. Natural resource inventories carried out by MAG's National Program for Agrarian Regionalization (PRONAREG) indicate that practically all of the soils in Ecuador that can support crop or livestock production have been occupied. Implicitly, the marginal benefits of additional colonization are meager.

Of the 14.99 million hectares covered by detailed inventories in the Costa and Sierra, PRONAREG has determined that 4.64 million hectares are suitable for crop production and can also be used to graze livestock. Approximately 70 percent of that land is relatively fertile, well drained, not highly erodible, and free of other serious natural limitations. It therefore offers excellent prospects for crop production, provided that water is available. Crops can be raised on the remaining 30 percent once measures to deal with erosion or drainage problems, for example, have been put in place (Southgate, 1990b, p. 83). An additional 2.56

million hectares in the western two thirds of Ecuador can be used for grazing. The remaining 7.79 million hectares should have permanent tree cover.

Current agricultural land use in the Costa and the Sierra is nearly identical to potential land use in the two regions. Adding the 4.05 million hectares of cropland and pasture west of the Andes (Table 4-1) to the 2.67 million hectares of agricultural land in the ten highland provinces (Chapter 5) yields 6.72 million hectares, which is 93 percent of the 7.20 million hectares that PRONAREG says can support either crop or livestock production.

By the same token, all the soils in the Oriente that lend themselves to agriculture, and many that do not, have already been settled by farmers and ranchers. In 1987, PRONAREG and other divisions of MAG completed an evaluation of 5.30 million hectares in northeastern Ecuador (MAG, 1987). The conclusion was reached that only 17 percent of the region (0.90 million hectares) was suitable for crop or livestock production and that forests should be maintained on the remaining 83 percent. When the evaluation was carried out, 1.10 million hectares had already been colonized (MAG, 1987). Settlement and deforestation have not stopped since then.

Evidence concerning the meager benefits of recent encroachment on Ecuador's tropical forests has been obtained in a 1991 survey of 179 agriculturalists living in the northwestern part of the country (Southgate, Chase, et al., 1992). Incomes among the sampled population were found to be very low. More than 33 percent of the colonists reported having negative cash flows for their respective farms and ranches. Only 10 percent of the sample took in more than $100 a hectare (at prevailing exchange rates) during the preceding twelve months. Most deforested land generated less than $25 per hectare per annum.

If the somewhat optimistic assumption is made that even this low income level can be sustained for twenty years (in the face of soil erosion, weed infestation, etc.), then the present value of the future benefits of land clearing is:

Value ($/ha)	Real discount rate (%)
310	5.0
255	7.5
210	10.0

These figures are consistent with another measure of the present value of the benefits of deforestation in the study area: local real estate values. Of the colonists who were willing to name a price, a third would sell out and move if offered $100 a hectare. Three quarters indicated that they would give up their holdings if offered $300 a hectare (Southgate, Chase, et al., 1992).

Deforestation's Opportunity Costs: The Commercial Portion

Low in and of themselves, the benefits of deforestation do not compare very well with one of the major opportunity costs of that land use change, which is the income earned by one who produces timber and other marketable forest commodities.

The net returns to timber production can be estimated on the basis of cost and yield data for two commercial species—pachaco (*Schizolobium parahyba*) and laurel (*Cordia alliodora*)—which have been planted on a few thousand hectares in northwestern Ecuador (Montenegro, 1987). Those net returns have been estimated for various scenarios regarding real interest rates, prices of standing timber, and the scarcity value of labor (Southgate, Chase, and Hanrahan, 1993).

Raising pachaco is more profitable, mainly because that species reaches a harvestable size eighteen years after planting, whereas a typical laurel rotation lasts twenty-three years. When wages are high by rural Ecuador's standards ($3.50 a day) and stumpage values and yields are low ($10 a cubic meter and 280 cubic meters per harvested hectare, respectively), the net returns for a pachaco rotation are as follows:

Net return ($/ha)	Real discount rate (%)
875	5.0
400	7.5
85	10.0

Of course, timber is not the only commodity obtained from forests. Neglected until very recently, collection of fruits, latex, and other commodities from tropical moist forests in Latin America is starting to generate substantial interest. One of the first economic studies of this activity was carried out by Peters, Gentry, and Mendelsohn (1989).

The approach they used to calculate "extractive" income was to multiply maximum sustainable yields of various products from a small plot near Iquitos, Peru, by local prices charged for those same products and then to subtract collection costs, which were assumed to equal 40 percent of sales revenues. The resulting estimate was a little over $400 a hectare.

If valid, this finding suggests that tree cover can be conserved and forest dwellers' economic welfare can be improved by establishing extractive reserves (to use the term currently being applied to primary forests used exclusively as a source of nontimber products). For example, multiplying what the typical colonist in lowland Ecuador has laid claim to (40 to 50 hectares) by $400 yields an annual income of $16,000 to $20,000, which is equivalent to per capita GDP in the world's richest countries.

Unfortunately for the sake of tropical forest conservation strategies based on the widespread establishment of extractive reserves, the study conducted by Peters, Gentry, and Mendelsohn (1989) has serious flaws. One criticism of their work has to do with an elemental truth of economics: downward-sloping demand curves. If there is a major expansion in the collection of nontimber products in the Amazon Basin, for example, prices are bound to fall, which will diminish annual net returns per hectare.

Another flaw of the study is that it rests on the absurdly unrealistic assumption that no products are lost between the point of collection and the point of sale. If, as one can reasonably suppose, half the output perishes, then net extractive earnings are only a sixth of what Peters, Gentry, and Mendelsohn (1989) say they are.

In a study carried out in the early 1990s, Grimes et al. (1993) found that indigenous groups in northeastern Ecuador use up to 95 percent of the species in local forests. They also determined that, at prices prevailing in local markets for foodstuffs, medicinal herbs, and other commodities, extracting nontimber products yields $35 to $232 per hectare in annual gross revenues.

In this chapter's analysis of the costs and benefits of agricultural land clearing, we suppose, conservatively, that annual extractive income is $20 a hectare. This means that a fifty-hectare holding (the size claimed by the average colonist in eastern or northwestern Ecuador) would yield an annual income of $1,000, which is nearly twice the legal minimum wage for rural areas. If land clearing fully interrupts collection of nontimber products for twenty years, then the present value of lost extractive income is:

Lost income ($/ha)	Real discount rate (%)
250	5.0
205	7.5
170	10.0

Deforestation's Opportunity Costs: The External Portion

The economic losses caused by agricultural land clearing do not end with diminished supplies of timber and other commodities. In a complete analysis of encroachment on natural habitats in eastern and northwestern Ecuador, environmental externalities must also be taken into account.

In all likelihood, the most serious environmental consequence of land use change in Ecuador has to do with diminished biodiversity. Myers (1988) has identified ten "hot spots" around the world, where tropical moist forests inhabited by large numbers of rare plants and animals are under severe threat. The Oriente is the center of one of these ten zones, the uplands of western Amazonia. The same investigator considers western Ecuador, where cumulative deforestation is at an advanced stage (Table 4-1), to be one of the world's three hottest hot spots, along with Madagascar and the remnants of rainforests along the Atlantic coast of Brazil.

Diminished biological diversity defies economic evaluation. For one thing, scientists cannot describe with much precision the linkage between habitat destruction and species extinction. In addition, placing a monetary value on biodiversity losses would be difficult even if documentation of flora and fauna extinction specifically related to tropical deforestation in Ecuador were available.

Regardless, it is safe to say that humankind as a whole would probably be willing to pay a considerable amount to maintain the existence of rare species in Ecuador. Similarly, the world's scientific community would probably pay large sums to make sure that the country's stockpile of genetic information is not lost.

Compared to evaluation of diminished biological diversity, economic analysis of tropical forests' contribution to climatic regulation is a little more tractable. Using a dynamic model of climate change, damages resulting from global warming (e.g., rising sea levels and reduced agricultural production), and economic responses to those damages, Nordhaus (1992) has estimated that the present value of economic losses caused by carbon emissions will grow from a little more than $5 a metric ton in the middle 1990s to $10 a ton in 2025. Peck and Teisberg (1992) have conducted similar research using a dynamic model with a more detailed description of the energy sector. They conclude that the cost associated with releasing a metric ton of carbon into the atmosphere already exceeds $10.

In a study of global warming damages caused by deforestation in the Brazilian Amazon, Schneider et al. (1991) use a conservative estimate of costs: $5 per ton. They also employ a measure of carbon released from deforested land originally published by Fearnside (1987): 120 metric tons a hectare. The resulting estimate of global warming costs associated with land clearing is

$5/ton × 120 tons/hectare = $600/hectare

Deforestation in Ecuador might have less of an impact on the world's climate than what is suggested by Schneider et al.'s (1991) analysis. Different from what tends to take place in Brazil, cleared land in, for example, the Oriente is not burned, principally because conditions are usually too wet. Recognizing this, we suggest that global warming damages in Ecuador might be as little as $300 a hectare.

Conclusions

Economists studying tropical deforestation tend to argue that the challenge is not to eliminate agricultural land clearing altogether but instead to find an efficient cumulative level of that activity. That level is found where the marginal environmental costs just equal the marginal net benefits of diminished tree cover (comprising global warming damage, disutility associated with reduced biological diversity, etc.). The marginal net benefits comprise the difference between agricultural income generated on newly cleared land and forestry income that has been forgone. As is indicated in Figure 4-2, efficient cumulative deforestation is almost certainly positive because, as a rule, marginal environmental costs are low and marginal net benefits are high if small areas have been cleared.

Figure 4-2 also illustrates what the benefit and cost estimates presented in Table 4-2 reveal about the inefficiency of recent land use change in tropical Ecuador. Current deforestation seems to make sense from an individual perspective only if interest rates are high. At 7.5 percent real interest, raising pachaco is much more profitable. At 5.0 percent real interest, even forgone extractive income is comparable to agricultural earnings.

When global warming damages and the costs of diminished biodiversity are taken into account, converting tropical forests into low-grade cropland or pasture cannot be justified on narrow economic grounds.

Causes of Agricultural Colonization

The preceding section's analysis begs a question. If tropical deforestation is inefficient, even for many of the individuals involved, why does it continue?

Market failure is part of the problem since agents of deforestation consider global warming, species extinction, and other environmental impacts as externalities. Furthermore, basic socioeconomic forces contribute substantially to encroachment on Ecuador's natural ecosystems. Income distribution has always been highly skewed and per capita GDP, which is not much more than $1,000, is stagnating. In spite of emigration and recent declines in fertility, the country's population is still growing by around 2 percent a year. Migration to agricultural frontiers in the northwest and the Oriente is a response to the human realities underlying these national statistics.

But deforestation cannot be attributed simply to mounting poverty and demographic pressure in the countryside. It is interesting to note, for example, that tree-covered land was being cleared very rapidly in Esmeraldas province during the 1970s even though its rural popula-

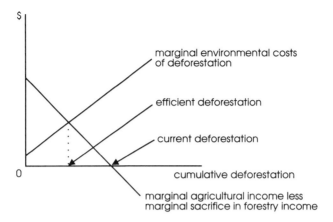

Figure 4-2 Efficient versus current deforestation.

Table 4-2 The Benefits and Commercial Opportunity Costs of Recent Tropical Deforestation in Ecuador

Real interest rate (%)	Agricultural income ($/ha)	Lost forestry income ($/ha)	Lost extractive income ($/ha)
5.0	310	875	250
7.5	255	400	205
10.0	210	85	170

tion actually declined between the 1974 and 1982 censuses (INEC, 1991).

Agricultural land clearing, it must be emphasized, has much to do with public policy. As is explained in the preceding chapter, Ecuadorian crop and livestock yields are low because too little has been invested in research and extension. As a result, encroachment on natural ecosystems is the only way for farmers and ranchers to respond to increasing commodity demands. In addition, reforestation and management of existing stands of trees are impeded by a general lack of long-term credit, caused by chronically high inflation.

Waste and misuse of forest resources are also brought about because governmental claims on those resources are not matched by administrative capacity in the public sector. The imbalance is illustrated by the ratio between lands set aside in parks and other reserves and the number of temporary and permanent guards: nearly 12,000 hectares per guard in 1991. Where the government cannot control what it claims, resources tend to be depleted by private individuals, who have no property rights. Among major examples of this behavior in "paper parks" and other areas formally designated as public properties are excessive fuelwood collection, poaching of wildlife, and illegal land clearing.

Depletion of public forests also occurs because private access is severely underpriced. For example, visitors have had to pay a nominal charge (generally less than $0.25 a person) to enter any park or nature reserve on the Ecuadorian mainland. Until early 1993, when fees were raised, loggers operating in those parts of the Forest Patrimony that are not included in a park or reserve had to pay only 1,350 sucres (equivalent to $0.70 at current exchange rates) for each cubic meter of harvested timber. Similarly, colonists claiming land in the Patrimony have had to pay nothing more than a small adjudication fee. For example, the minimum per-hectare fee, which is applied in most cases, was raised from 1,000 to 10,000 sucres in 1989, when the dollar was worth about 500 sucres.

Clearly, these charges do not begin to reflect the opportunity costs of resource use associated with ecotourism, logging, and agricultural settlement. As a result, resources are wasted and mismanaged and encroachment on natural ecosystems is excessive. The funds needed to administer protected areas are chronically scarce; in 1991, management and development expenditures in continental Ecuador's fourteen sites amounted to less than a dollar a hectare. In addition, the Ecuadorian wood products industry is, with a few exceptions, characterized much more by resource mining than it is by resource management. Partly because timber is priced cheaply, logging operations are wasteful and inefficient, processing involves low levels of wood recovery, and very few trees have been planted (Southgate, Stewart et al., 1993). Finally, the adjudication fee has had little or no impact on the behavior of colonists, most of whom still sense that partial deforestation is needed to demonstrate informal tenure to local IERAC agents and others (Southgate, Chase, et al., 1992).

Resolving the imbalance between excessive governmental claims on forest resources and the public sector's limited capacity for resource management is probably the most important step that can be taken to arrest tropical deforestation. More than anything else, this involves granting legal recognition to the de facto rights of indigenous communities, colonists, and others who have already occupied most of the Forest Patrimony. Among others, Pichón and Bilsborrow (1992) have documented that there is a positive linkage between secure property rights and natural resource conservation in areas undergoing or threatened by deforestation.

However, strengthening private and community tenure in tree-covered land is not sufficient to ensure sustainable development of forest resources. Policies that currently depress domestic timber prices will have to be reformed in order to give land owners an incentive to reforest and to manage existing stands of trees (Southgate, Stewart, et al., 1993).

Overvaluation of the sucre has caused producers of timber sold in international markets to receive less for their output and nominal taxes have been assessed on wood exports. In addition, bans, quotas, and other restrictions applied to overseas shipments have had a significant impact on domestic timber prices.

With passage of the Export Facilitation Law of 1992, minimum processing requirements for commodity exports were removed and it became more difficult for government officials to cite a domestic raw materials shortage as an excuse for interfering with international trade. However, the same legislation also states that indigenous flora and fauna in danger of extinction are not exportable.

Officials have seized on the latter clause to continue regulating the forest product trade. In particular, a list of more than thirty species that cannot be exported, which was presented by MAG in December 1990, remains in effect. The list includes types of trees that are relatively abundant and consumed in great quantities domestically, like chanul (*Humiriastrum procerum*) and sande (*Brosimunn utile*).

The consequences, and purposes, of restrictions on timber exports were made very clear in Ecuador in 1989, when an Italian firm was offering to buy eucalyptus (*Eucalyptus globulus*) for $15 or more per cubic meter. The domestic price at that time was $2 to $4 a cubic meter. Some Ecuadorian environmentalists saw fit to oppose this trade. From the standpoint of forest conservation, their position is difficult to understand because eucalyptus coppices immediately after cutting. All that they accomplished was to serve the interests of the domestic wood products industry, which supported the export ban enthusiastically so that raw materials would stay artificially cheap.

Finally, it must be recognized that, even if Ecuadorians with secure property rights faced strong price incentives to grow timber, their ability to do so would be severely limited because the forestry sector's technological base is weak. Aside from one or two companies undertaking species trials, private sector forestry research is negligible. Government research is correspondingly modest and, for all intents and purposes, there is no forestry extension. These deficiencies will have to be remedied if Ecuador's potential for timber resource development is to be realized and if agricultural colonists and other agents of deforestation are to perceive that their activities involve a significant opportunity cost.

Summary and Conclusions

By Latin American standards, Ecuadorians are converting natural ecosystems into farms and ranches very rapidly. Between 1982 and 1987, the area dedicated to crop and livestock production increased by 2 percent a year. The only country in the Western Hemisphere experiencing higher growth was Surinam, where the initial base of cropland and pasture was very small (Table 4-3).

Ecuador has reached the point where geographic expansion of the agricultural economy is no longer a viable response to mounting commodity demands. Nearly all the land in the Costa, Sierra, and Oriente that can be used for crop or livestock production has already been occupied by farmers and ranchers. Recent agricultural colonization has been a resounding economic failure for most of the people directly involved, as surveys carried out in northeastern and northwestern Ecuador show (Pichón and Bilsborrow, 1992; Southgate, Chase, et al., 1992).

It is also a failure as far as society is concerned. Agricultural income in recently cleared areas compares poorly with the opportunity costs of agricultural land clearing. The earnings that could be generated by commercial forestry and the collection of nontimber products are forgone. In addition, deforestation results in species extinction, global warming, and other adverse environmental consequences.

Inefficient land clearing persists largely because agricultural colonists and other agents of deforestation do not internalize all the costs of their actions. Obviously, they regard environmental impacts as externalities. Furthermore, the destruction of timber and other resources is ignored because frontier tenurial arrangements cause colonists to value land only after it has been cleared and because trade restrictions and other policies drive down stumpage prices.

To save Ecuador's tropical forests, an integrated reform of public policy is required. The property rights of those who use forest resources must be strengthened and the prices they receive for timber and other commodities must be allowed to rise to international levels. In addition, research and extension must be improved so that individuals and firms

Table 4–3 Five Latin American Countries with the Highest Rates of Agricultural Frontier Expansion, 1982 to 1987

	Annual growth in cropland and pasture (%)
Surinam	3.2
Ecuador	2.0
Belize	1.2
Costa Rica	1.1
Paraguay	1.0

Source: Southgate (1991).

will respond to incentives for timber production by managing forest resources instead of mining them.

Fundamental changes in policy are needed to induce patterns of development that make sense now that the era of geographic expansion of the agricultural economy is coming to an end.

5

Farmland Degradation

By no means is policy-induced environmental degradation confined to agricultural frontiers. Where crop and livestock production are already well established, in Ecuador and elsewhere in Latin America, government policies discourage the conservation of land resources.

Soil loss is the primary focus of this chapter. Farmers' and ranchers' contributions to soil displacement are examined and erosion's agricultural impacts are surveyed. (Off-farm impacts are addressed in the next chapter.) Since data needed to estimate linkages between erosion and agricultural production are lacking in Ecuador, this chapter's causal analysis of farmland degradation is largely inferential. Our principal argument is that policy-induced distortions in real estate and financial markets discourage the adoption of land improvements, generally, and conservation measures, specifically. The burden of these distortions weighs heavily on small holders, who tend to be concentrated on the country's more fragile lands.

Soil Erosion and Its Origins

In terms of being a medium for crop and livestock production, few of Ecuador's soils are entirely free of natural limitations. In both the Sierra and the Costa, careless irrigation can lead to waterlogging, salinization, or both. Also, soils that are alkaline or have a high clay content are easily compacted by heavy farm machinery.

The most serious threat to agriculture's land resource base has to do with erosion. Soils are easily displaced, slopes are pronounced, or both in many places. Rainfall erosivity varies from place to place. For example, heavy storms are not very frequent in the central and southern Sierra (Harden, 1991). By contrast, Moya and Peñafiel (1989) have conducted experiments that demonstrate that rainfall erosivity is fairly high in the northern highlands. Taking these conditions into account, de Noni and Trujillo (1986, p. 6) conclude that the potential for soil loss is great in 47.9 percent of continental Ecuador. Vulnerable areas are not confined to the Andes and their outer flanks. Coastal hills, the northwestern part of the country, and much of the Oriente are also at risk.

Natural erosion, as opposed to what humans induce, is important in many parts of the country. For example, some of the sediments working their way through highland watersheds, much to the detriment of water resource development projects (Chapter 6), are scoured out of the banks of rivers and streams, which will be cutting their way through narrow gorges for a long time to come. To processes such as these must be added the soil displacement resulting from human activity. Building a road along a mountainside, for example, can cause thousands of tons of sediments to change their position, and rock and gravel extraction creates major downstream impacts in some places. In addition, removing natural vegetation to make way for agriculture leaves land more exposed to wind and rain, which carry off a large amount of soil.

Soil Loss Due to Livestock Grazing

The origins of contemporary livestock production, and its environmental impacts, can be traced back to the Conquest. Soon after arriving in the Andes, Spaniards began to introduce cows, horses, burros, goats, and sheep. Sheep, in particular, quickly displaced llamas, which were indigenous to the region. Both species yield similar wool and flourish on the coarse forage of the *páramos* (grasslands in the high Andes). But whereas native camelids are pad-footed, sheep (and goats) have sharp, cloven hooves, which fracture the soil. Better than llamas at digging, livestock brought over from Europe like to nibble on stems and roots beneath the soil surface. As a consequence, they can do more damage to *páramo* grasses, which tend to be poorly adapted to heavy grazing and trampling by animals.

Herders, themselves, sometimes do complementary damage. Long ago, they learned that periodically burning old grass stems promotes the emergence of new growth, which is more palatable to livestock. To this day, the driest months of the year, when fires are relatively easy to set, are anxiously awaited, even though winds carry off mineral-rich ashes. In addition, soil loss, driven by wind and rain, often follows fires.

Domesticated animals whose ancestors were brought from Europe have also become plentiful at lower elevations. In recent decades, pasture area has increased considerably in the Costa and Oriente at the expense of tropical forests (Chapter 4). The damage to soils and natural vegetation done by livestock seems to be especially severe in dry forests, both on the continent and in the Galápagos, where endemic species are being driven to extinction due to the introduction of exotic flora and fauna (Chapter 9).

Erosion Due to Crop Production

Along with European livestock, Spanish colonists settling in what was to become Ecuador brought in new crops (e.g., wheat and barley), the raising of which leaves land highly exposed to the elements. New technology for crop production was also introduced and conservation measures used before the Conquest (e.g., terraces) were abandoned.

Until recent times, plowing and related tillage practices probably did not cause too much damage because crop production was concentrated in Andean valley bottoms and other places where erosion risks are relatively low. Also, rotations of corn, small grains, legumes, and forage crops were traditionally employed to keep soil in place and to maintain fertility. However, the situation has changed in recent decades, both because more erodible land has been planted to crops and because mechanized tillage has become more widespread.

Cropland has increased by nearly 30 percent in the Costa and Oriente during the past two decades (Table 4-1). Erosion problems have arisen because some of the newly cultivated lands are characterized by thin soils or steep slopes. The area planted to annual and perennial crops in the Sierra has declined by a third since the early 1970s, for reasons that are explained later in this chapter. However, many of the small farms (*minifundios*) established on environmentally fragile Andean hillsides as a result of the 1964 Agrarian Reform (Commander and Peek, 1986) continue to operate.

Until recent times, machinery powered by diesel fuel and gasoline was used only on larger holdings. Although animal traction remains widespread, small farmers' access to tractors and other machinery has been promoted by the National Mechanization Program (PRONAMEC). Established in the 1970s, the agency has run seventeen centers around the country since 1982. In spite of extension agents' recommendation to plow along contours, PRONAMEC operators frequently run their machinery up and down slopes (F. Maldonado, personal communication, 1992). They do so to avoid tipping over and also for the sake of convenience (the length from top to bottom of the typical highland field is much greater than its cross-slope width). In any event, plowing with tractors up and down slopes can lead to the formation of gullies, down which large volumes of soil can travel.

On-Farm Consequences of Soil Loss

Due to overgrazing and depletive crop production, soil loss has become severe in many parts of Ecuador. Erosion is either "active" or in the process of becoming so in 12.3 percent of the country (de Noni and Trujillo, 1986, p. 6). Most of the land they say is currently undergoing degradation is in the Andes. Soil loss is also heavy in coastal hills and on agricultural land carved out of tropical forests in northwestern and eastern Ecuador during the past few decades.

Evidence concerning actual erosion rates, and their determinants, in different parts of Ecuador is extremely spotty. Studies have only been carried out in a few isolated locations. Referring to research conducted in the 1970s, Moya and Peñafiel (1989) contend that fields with a moderate (i.e., 15 to 20 percent) slope lose 83.5 metric tons of soil per hectare each year if they are devoid of plant cover. Using rainfall simulation equipment, Harden (1991) found that erosion is especially pronounced when a heavy storm hits right after land has been tilled (just before planting). Lamenting the lack of comprehensive research on the subject, Soria (1990) notes that per-hectare erosion rates in the Sierra range from 3 to 671 metric tons a year, the latter figure having been recorded on fields with 33 percent slope that are planted continuously to corn.

Without a doubt, erosion of this magnitude causes land productivity to decline. For a few years, output levels can be maintained by applying more fertilizer and through other means. But after cumulative soil loss passes a certain threshold, raising crops often becomes economically infeasible. For example, *cangahua* (a hard-packed material of volcanic origin) has either become exposed or lies just under the soil surface in 35 percent of the central and northern highlands (Caujolle-Gazet and Luzuriaga, 1986, p. 59). Once it has been uncovered, as a result of cumulative erosion, land abandonment is often a farmer's best economic option (G. del Posso, personal communication, 1992).

There is no empirical evidence on the relationship between soil loss and crop yields in Ecuador. However, it seems clear that deteriorating soil quality has caused some cropland to be abandoned, particularly in the Sierra.

Why Farmers Allow Their Land to Deteriorate

Alarmed about erosion and related renewable resource degradation, donor agencies and national governments have implemented soil conservation and watershed management projects throughout Africa, Asia, and Latin America. A major thrust of these projects has been to communicate information on erosion control techniques, agroforestry systems, and other measures to farmers and ranchers. Presumably, the target group wants to maintain the quality of its land but lacks the required know-how.

While valid in many settings, this approach neglects the various ways that public policies discourage efficient patterns of land use. Misguided policies also cause farmers not to adopt soil conservation measures.

Throughout Latin America, extensive cattle ranching on land well suited to intensive crop production is strongly condemned, particularly where fragile lands nearby are divided among small agricultural holdings. Precisely this sort of juxtaposition is considered by many to be a primary symptom of a highly flawed socioeconomic order in the Ecuadorian Sierra. Large dairy farms are dominant in valley bottoms, where soils are rich and water plentiful. Meanwhile, small farmers populate the surrounding hills, where erosion risks are often severe.

Of course, it is erroneous to think that grazing livestock on prime farmland always amounts to resource under-utilization. As is stressed later in this chapter, prices of wheat and other basic grains, which used to be the mainstay of the highland agricultural economy, have declined steadily in recent times. Many Andean valleys now hold a comparative advantage in dairy production.

Where it arises, an inefficient mix of under-utilized prime farmland and over-worked, erodible hillsides has much to do with public policies. Livestock and crop production, it must be remembered, are largely exempt from taxation, because of favorable provisions in the tax code and also because it is fairly easy for farmers to avoid taxes. Furthermore, real estate taxes are low in rural areas and, up until the early 1990s, collection was often sporadic. As is the case in other parts of the world, all this beneficial treatment is especially valuable for rich people, who have larger tax liabilities. They are also eager to invest in real estate in order to hedge against inflation, which of course results from misguided macroeconomic policies.

Prime farmland, then, often ends up in the hands of individuals trying to avoid taxes and to maintain portfolio values. Those same individuals are often not particularly adept at or even interested in farming. Poorer individuals, whose relative ability to compete for high quality real estate is diminished by the tax code and chronic inflation, end up with inferior parcels.

Once relegated to hillsides and other fragile lands, small farmers have weak incentives to adopt erosion control practices and other land improvements because of the public policies described in Chapter 3. Particularly in agriculturally marginal areas, property rights are insecure and financial intermediation is impeded because there is no way to verify title accurately and cheaply and because agrarian reform legislation has created transactions costs and expropriation risks. In and of itself, tenure insecurity makes land owners less willing to make investments that yield benefits only after several years have passed. Furthermore, financial sector repression causes the credit needed for soil conservation to be artificially scarce and makes it difficult to capture the long-term benefits of land improvements, in the form of higher real estate values.

No group of poor farmers has weaker property rights than the members of agrarian communes (*comunas* and *cooperativas*). Passed in 1937, during a fit of admiration for pre-Columbian group tenure (as perceived by romantics) and Soviet agricultural collectives, the Commune Law stipulates that commune members possess usufructuary privileges in the land where they plant crops or graze livestock. They cannot transfer interests in resources, either through sale or inheritance. Neither can a commune be divided among private holdings.

These restrictions make members of *comunas* and *cooperativas* reluctant to make productivity-enhancing investments, since they doubt that they will be able to internalize those investments' full benefits. In addition, the prohibition on transferring group properties makes it impossible to use those properties as collateral for loans needed to finance land improvements.

A study recently carried out in the central and northern Sierra suggests that investment in erosion control, improved water management, and other measures that boost yields is well below efficient levels on most *comunas* and *cooperativas*. In particular, Camacho and Navas (1991) found that the ratio of crop and livestock output to land area in group holdings is less than a third of the ratio for privately owned farms, be they large or small.

The most damaging consequence of agrarian reform in several Latin American countries has been to declare large tracts of the countryside as *ejidos*, *comunas*, and *cooperativas*. More than two thirds of the agricultural land in the four jurisdictions that Camacho and Navas (1991) surveyed, for example, are in group holdings. For all intents and purposes, those areas have been removed from the domain of formal real estate and financial markets. Implicitly recognizing the burden that this creates for small farmers and the natural resources on which they depend, several Latin American countries are giving agrarian reform laws a thorough overhaul (Hendrix, 1993). One of them is Mexico, where *ejidos* have been sacrosanct for generations (Fraser, 1991).

Land Degradation and Declining Commodity Prices

In and of themselves, the policy-induced real estate and financial market distortions described in Chapter 3 and the preceding section take a serious toll. The damage has been made worse in recent years because agricultural commodity prices have been falling.

In large part, the decline has been caused by the application of economic development strategies predicated on import substitution and industrialization (Chapter 3). In addition, Ecuador has not been able to insulate itself from strong trends toward lower prices in international markets. Due to subsidies in rich countries for agricultural production and exports and also to yield increases, real grain prices declined by more than a third during the 1980s (FAO, 1992, p. 25).

In highland Ecuador, where soil erosion is particularly severe, falling crop prices have also had much to do with infrastructure development. As recently as the 1960s, most Andean valleys had rudimentary road and rail links with the outside world. Almost by necessity, those places were traditionally self-sufficient in the production of food, generally, and staple grains, specifically. Road construction, which began in earnest in the 1960s and accelerated in the 1970s (when Ecuador became a petroleum exporter), exposed highland farmers to unprecedented internal and external competition. Rice, plantains, and other tropical crops were brought up from the Costa in large quantities. The downward trend in prices for barley, rye, and, most important, wheat was further accelerated because the Sierra found itself in the novel and uncomfortable position of competing with North America, Australia, and Argentina.

Ecuador's Andean farmers have responded as their counterparts in neighboring countries have done to the diminishing profitability of grain production. They are growing more potatoes and beans (Ramos and Acosta, 1991). In addition, output of cut flowers, asparagus, and other specialty crops, which are sold in North America and other markets around the world, has risen.

There has also been a major switch to livestock production. Estrada (1993) points out that the conversion of cropland to pasture, which has affected millions of hectares from Bolivia to Colombia, is explained primarily by the fact that the prices of milk, meat, and other livestock products have risen relative to the market values of crops. This trend has to do with fundamental market forces (e.g., enhanced demand for livestock products caused by income growth) as well as public policy (e.g., controls on basic grain prices and limited protection for dairy processors).

For two reasons, the dimensions of land use change in the Ecuadorian Sierra are difficult to gauge. First, pasture area is reported only at the provincial level and ranchers have cleared hundreds of thousands of tropical forests in the western, lowland portions of several highland provinces during the past quarter century. Second, some of the land at higher elevations that is classified as pasture is actually degraded cropland on which cattle and sheep occasionally browse. But in spite of data limitations, it is clear that cropland has declined and the area devoted to livestock production increased between the early 1970s and late 1980s (Table 5-1).

Table 5-1 Agricultural Land Use Trends in Highland Ecuador

Land use	1972–73 (ha)	1988–89 (ha)	Change (%)
Cropland	503,000	325,000	−35
Pasture	1,024,000	2,349,000	129

Source: MAG and SEAN-INEC reports.

Notwithstanding the growing importance of cattle and sheep ranching, large declines in grain production were bound to stimulate a certain amount of depreciation of fixed assets in the Andean agricultural economy. Soil erosion as a prelude to the abandonment of cropland can be viewed as an example of this behavior.

Of course, decisions about what to do with real estate should not be driven exclusively by reduced profitability of what has been its predominant use. Other purposes to which land can be put, both now and in the future, need to be examined. For example, the possibility that some alternative use will be remunerative a few years from now should be considered before allowing soils to be lost.

Unfortunately, the policy-induced distortions in financial and land markets that are analyzed in this volume have provoked an intertemporally inefficient response to deteriorating terms of trade for agriculture. Credit shortages have discouraged land users from considering the future consequences of current resource use and management. In addition, attenuation of private land rights has caused owners to doubt that they will be able to capture the long-term benefits of conservation.

Another consequence of declining crop prices has been to encourage people to seek jobs outside of agriculture, primarily through rural-to-urban migration. Commander and Peek (1986) point out that improved employment opportunities have been drawing people off Sierran farms since the 1960s, at least. Census returns demonstrate that hundreds of thousands of peasants saw fit to relocate during the past quarter century (Chapter 2).

Emigration from the countryside need not be a cause of land degradation, provided that real estate and financial markets are allowed to function efficiently. However, policy-induced distortions in those markets discourage families from using and managing their land efficiently before moving to cities and towns. Since credit is scarce and real estate values do not reflect very well the returns to an investment in soil conservation, the best option for many families is to live off the proceeds of soil mining prior to resettling. For example, relocation often begins with a household's teenage and adult men circulating periodically to urban areas in order to work on construction projects or in other jobs. Wives, children, and other dependents can stay on farms and help support themselves with the produce they raise, often by farming in ways that exhaust renewable resources. Once depletion has run its course, land can finally be abandoned.

New Directions in Soil Conservation Policy

In Ecuador and throughout the developing world, dissemination of improved technology for erosion control is the principal feature of soil conservation programs and projects. Various benefits are claimed for those initiatives, including increased agricultural production, reduced sedimentation and flooding in downstream areas, and diminished rural-to-urban migration.

To be sure, the need for technology transfer is great. Working in dozens of Andean communities, Project CARE and its partners in MAG have demonstrated what can be accomplished when extension agents work closely with farmers to deal in an integrated fashion with agricultural production and fuelwood availability problems brought about by soil loss and deforestation. It is also likely that some beneficiaries of the CARE-MAG program have found that prospects in the countryside have improved enough to persuade them not to relocate to an urban slum.

This chapter's analysis, general though it is, suggests that technology transfer must be complemented with policy reform if farmland degradation is to be arrested. Inflation has to be controlled and tax codes must be changed in order to remove the disadvantages small operators now face when competing for prime farmland. Strengthening property rights, through the modernization of cadasters and registries and through fundamental changes in agrarian reform legislation, will allow rural real estate and financial markets to provide incentives for land improvement, generally, and soil conservation, specifically. Macroeconomic and sectoral policies responsible for excessive declines in commodity prices also have to be eliminated.

Major demographic changes are taking place in the developing world. Ecuador, for example, is being transformed from a rural nation to one in which most people live in cities and towns and work outside of production agriculture. Major investments in human capital are needed to ease this transition for the individuals most directly involved. Otherwise, too many people will find that their best alternative is to practice depletive agriculture in environmentally fragile areas. In addition, removal of the policy-induced distortions in commodity and input markets criticized in this chapter is essential if individuals are not to waste renewable resources as they make their way from farming to other sectors of the economy.

6

Waste and Misallocation of Water Resources

By any standard, water supplies are abundant in Ecuador. There are several major river basins and aquifers in the country that, taken together, are more than enough to satisfy all conceivable needs. Nevertheless, competition for water is increasing, with demand outstripping supply in drier areas. Shortages are partly the result of watershed deterioration and pollution. However, inefficient and inequitable use of water has also been caused by government policies.

For twenty years, the Ecuadorian Institute of Water Resources (INERHI) has held primary responsibility for implementing those policies. But its broad mandate to supervise water resource development throughout the country has been resisted successfully by various national, regional, and local authorities intent on pursuing their own agendas. INERHI, then, has chosen to concentrate on building and operating its own costly irrigation projects.

With very few exceptions, public sector initiatives to deliver water to farmers have failed to meet expectations. The benefits of irrigation are generally well below its costs and tend to be captured by a few relatively affluent individuals. Along with the deficiencies of INERHI and its partner institutions, the policy of selling water at a small fraction of its cost causes waste and misallocation.

Since farmers use far more water than any other group in the country, irrigation is a major focus of this chapter. As is demonstrated in brief reviews of hydroelectricity and potable water development, though, inefficiency and inequity resulting from underpricing is a ubiquitous problem in Ecuador. We also point out that spending money on subsidies has left the government with inadequate means to manage watersheds and to control pollution. Ultimately, this could be the most damaging consequence of the current policy environment.

Water Resources for Agriculture

Ecuador is richly endowed with water resources, which are reasonably well distributed around the country. Aquifers have been developed for municipal systems and irrigation in both the Sierra and Costa. Total surface discharge toward the Amazon River is between 210 and 370 billion cubic meters a year and annual surface flow into the Pacific Ocean is 80 to 140 billion cubic meters (Delavaud, 1982, pp. 12–13). Most of Ecuador's eighty-four river basins are west of the continental divide.

To put resource endowments into perspective, consider that, if the lower limit of annual discharge into the Pacific were spread uniformly over the 1.58 million hectares used to produce crops in the Sierra and Costa (Tables 4-1 and 5-1), every hectare would receive more than 50,000 cubic meters a year. This would be equivalent to covering all cropland outside the Oriente with more than 5 meters of water annually. By contrast, per-hectare water requirements in INERHI systems averaged a little less than 20,000 cubic meters in 1987.

Not even the most ambitious irrigation planner would recommend diverting all Ecuador's surface waters onto agricultural land. Obviously, alternative uses of the resource need to be considered. In addition, farmers' and ranchers' demand for water is not strong in all parts of the country. Taking into account temperature, elevation, and plant and soil characteristics, Delavaud (1982, pp. 17–20) determined that agricultural water projects are appropriate at higher altitudes and also in the central and southern Costa.

For the most part, irrigation is used in the Sierra to supplement natural precipitation. Relatively low temperatures limit crop production, and therefore water demand, in July, August, and September, which are dry months. The beginning of the principal agricultural season coincides with the rains that start in October and reach a peak in November. Another peak in precipitation, occurring in April, heralds the initiation of a secondary agricultural season, which comes to a close with the return of dry weather. By and large, the design of highland systems reflects the need to make up for periodic shortfalls in precipitation during the nine-month wet season. Canals for stream diversion are commonplace while reservoirs for storing water are rare.

Different from the Sierra's bimodal distribution, rainfall in the central and southern Costa has a single peak each year, which is typical of what one finds at lower elevations throughout the tropics. The wet season begins in December, usually quite abruptly after six months of very dry weather. Precipitation rises through March and then trails off dramatically.

During the wet season, most farmers west of the Andes have very little need for water to supplement what falls from the sky. Indeed, flooding is often a serious problem during the first few months of the calendar year. The principal benefit of irrigation development, then, is to allow a second or third crop to be produced after rainfall has peaked. Construction of reservoir capcity is essential for capturing this benefit since water flow in streams originating in the Costa (especially in the Guayas River and its tributaries) is, like precipitation, characterized by marked seasonal variation.

The Water Law and INERHI

Legal interests in Ecuador's water resources have changed greatly in recent decades. An old tradition of private ownership, involving the right to buy, sell, and even rent water separately from agricultural land, was codified in legislation passed in 1936. The tradition survived through the 1960s, although owners' prerogatives were reduced somewhat over time.

Private rights were extinguished by the 1972 Water Law, which nationalized all of Ecuador's water resources. Former owners of water were given usufructuary privileges. However, those privileges were not transferable and were subject to forfeiture if the state determined that water was being used inefficiently.

The 1972 law gives INERHI, which was created in 1966, a great deal of responsibility. The first of its three primary functions is to plan, administer, and regulate, as the ultimate authority of the government, the use of all water for any purpose, be it public or private, rural or urban. INERHI's second function is to plan and administer, again as the state's maximum authority, all irrigation, drainage, and flood control activities in the country. Third, it is charged with designing, constructing, and operating agricultural water systems (Cruz, Orquera, and Salazar, 1986).

Strict interpretation of the Water Law seems to give INERHI supervisory control over a number of public agencies. The most important at the national level is the Ecuadorian Institute for Electrification (INECEL), which has pursued the ambitious program of hydroelectricity development described later in this chapter. Aside from a number of provincial and local governments, there are four regional authorities involved in water resource development: the Commission for the Study of the Development of the Guayas River Basin (CEDEGE), the Center for the Rehabilitation of Manabí (CRM), the Center for the Economic Recovery of Azuay, Cañar, and Morona Santiago (CREA), and the Regional Program for the Development of Southern Ecuador (PREDESUR).

Each of these agencies carries political clout and is reluctant to let any outsider encroach on its prerogatives. By contrast, INERHI is a relatively weak bureaucracy. From 1985 through 1990, it had a new executive director, always a political appointee, every twelve months or so. In addition, the institute has suffered from labor problems. In 1988, for example, 1,357 unionized workers went on strike for three months, which among other things closed down the Pisque Project during the height of the dry season (Figure 6-1).

In light of political realities, it is hardly surprising that INERHI tends

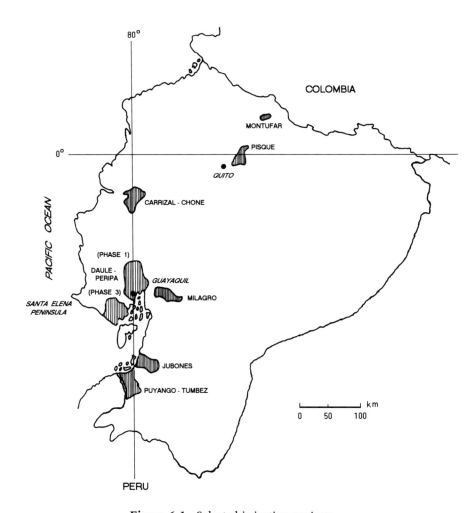

Figure 6-1 Selected irrigation projects.

to neglect its first two primary functions. Of the institute's total budget during the period 1985 through 1988, only 3.3 percent was devoted to supervising water resource development. The administrative unit responsible for this task is subordinate to the Technical Directorate, which builds and operates INERHI's irrigation systems. In addition, the practice of earmarking funds effectively prevents the institute from overseeing all the country's irrigation, drainage, and flood control activities. For example, it receives tax monies from the Plan Loja that must be spent on irrigation in Loja province. Similarly, half of the misnamed National Fund for Irrigation and Drainage (FONARYD), which is supported with a small tax on Central Bank credits, must be directed to the Jubones Project (Figure 6-1). Of all the money spent by INERHI from 1985 through 1988, a mere 0.1 percent was devoted to the second primary function mandated by the Water Law, which is to oversee national irrigation development (Whitaker and Alzamora, 1990a, p. 164).

INERHI, then, is left to concentrate on the third mission, which is to design, construct, and operate its own projects. These activities absorb 87 percent of the institute's budget (Whitaker and Alzamora, 1990a, p. 164).

Irrigation Projects

Since the 1970s, INERHI and other agencies have attempted to implement an ambitious national strategy to increase agricultural production and rural income by delivering more water to farmers. Large sums of money have already been spent and projects that have not yet come on line will be very costly. Unfortunately, the returns on that investment have been disappointing, in terms of both efficiency and equity. More than anything else, waste and misallocation have to do with the policy of selling water for much less than either its cost or its value to farmers.

Major Public Systems

As of 1989, there was a little more than 115,000 hectares in public irrigation projects, with combined capital costs of $251.0 million, in 1988 U.S. dollars. (Unless indicated otherwise, all values reported in this chapter are given in 1988 dollars.) Of the total area, around 70 percent was accounted for by INERHI's thirty-five projects and nearly 15 percent was in systems operated by El Oro's provincial government. Private and community systems had a combined area of 330,000 hectares, which means that, all together, a quarter of all cropland in the Sierra and Costa was benefiting from irrigation (Whitaker and Alzamora, 1990a, pp. 167–172 and p. 177).

If all projects currently under construction or in the planning stage are executed, irrigated area will increase substantially. New INERHI and

CEDEGE initiatives will benefit 62,000 and 33,000 hectares, respectively, and small systems being built by other agencies will deliver water to 13,000 hectares. If projects now being designed or given serious study are put into operation, another 294,000 hectares will be irrigated (Table 6-1).

The costs of all this infrastructure development are considerable. Capital spending on the projects now under construction (Table 6-1) amounts to $1.1 billion, which is nearly equivalent to a tenth of Ecuador's outstanding foreign debt. The proposed systems listed in Table 6-1 will require that an additional $1.1 billion be invested (Whitaker and Alzamora, 1990a, p. 177).

To infrastructure expenses must be added operation and maintenance costs, which will be high. In the first phase of the Santa Elena Project, for example, water will have to be lifted 73 meters before being released into the canal system. The vertical distance to the intakes to the second phase of the project and the Carrizal-Chone Project is 75 meters higher. As is indicated in Table 6-1, a major purpose of Daule-Peripa (II) is to supply the power needed for pumping. Also to be taken into account are spending on extension and other services provided by INERHI and other governmental agencies, institutional development expenses, and the costs of central administration and management.

Pricing Policy

Formal policy statements say that the costs of delivering irrigation water through public systems should be paid entirely by the beneficiaries of those systems. "Basic" and "volumetric" tariffs are supposed to cover capital and current costs. In practice, neither comes close to serving its respective purpose.

Subsidization of capital costs is especially extreme. For example, INERHI has forgiven a quarter of all spending on infrastructure and it has not revalued costs to account for inflation. The recovery period has been seventy-five years, during which time no interest has been charged. Under the circumstances, the basic tariff is estimated to have been only $1.57 per irrigable hectare for the thirty-five INERHI systems in operation in 1989. The actual cost of amortizing the investments in the same projects, assuming a forty-year project life and a 6 percent discount rate, averaged $144.73 per irrigable hectare, which means that nearly 99 percent of all capital costs were being paid by the state (Table 6-2).

An attempt is being made to reduce capital cost subsidies. Recovery of interest expenses is not being contemplated at present. In addition, there has been no change in the old practice of charging an average basic tariff for all systems, which causes more efficient farmers to subsidize the less efficient as well as misallocation of irrigation water and other inputs. But in 1993, INERHI reduced the amortization period to fifty years and

Table 6-1 Irrigation Development in Ecuador

Project name	Executing agency	Description
Systems under Construction		
Daule-Peripa (I)	CEDEGE	A dam was completed on the Daule River, upstream from Guayaquil, in 1989; controlling stream flow has made potable water supplies more reliable in the port city, reduced flooding, and made navigation easier; in addition, 17,000 hectares west of the river will be irrigated.
Santa Elena (I)	CEDEGE	Once the canal system is finished, withdrawals from the Daule-Peripa Dam will increase potable water availability and 16,000 hectares will be irrigated in the semiarid peninsula southwest of Guayaquil.
Various	INERHI	62,000 hectares will be irrigated in different parts of the country.
Various	Other	13,000 hectares will be irrigated in different parts of the country.
Projects Being Designed or Given Serious Study		
Carrizal-Chone	CRM	18,000 hectares in Manabí province will be irrigated, partially with water withdrawn from the Daule-Peripa Dam.
Daule-Peripa (II)	CEDEGE	130 megawatts of generating capacity will be installed on the Daule-Peripa Dam, largely to supply energy for Carrizal-Chone and Santa Elena projects.
Daule-Peripa (III)	CEDEGE	33,000 hectares east of the Daule River will be irrigated.
Puyango-Tumbez	PREDESUR	76,000 hectares will be irrigated in Loja province.
Santa Elena (II)	CEDEGE	27,000 more hectares will be irrigated and potable water supplies will be improved in the peninsula.

(*Continued*)

Table 6-1 (Continued)

Project name	Executing agency	Description
Various	INERHI	132,000 hectares will be irrigated in different parts of the country.
Various	Other	8,000 hectares will be irrigated in different parts of the country.

Source: From Whitaker and Alzamora (1990a).

Table 6-2 Costs, Tariffs, and Subsidies in INERHI Irrigation Projects ($/ha)

	Yearly cost	Annual tariff	Annual subsidy
Infrastructure amortization	144.73	1.57	143.16
Operations and maintenance	15.70	7.02[a]	8.68
Total	160.43	8.59	151.84

Source: Authors' calculations based on data reported in Whitaker and Alzamora (1990a), pp. 164, 180.
[a] Volumetric tariff for land planted to potatoes, which are a water-intensive crop.

started including 100 percent of all infrastructure expenses (valued in constant, as opposed to historical, prices) in the tariff base. Although government still covers more than 95 percent of all capital costs, some farmers have resisted the changes because the basic tariff has gone up by 342 percent.

As is the case with the basic tariff, an average volumetric tariff is charged across the entire INERHI system, which results in resource misallocation. In addition, major subsidization of operation and maintenance costs occurs because INERHI does not include various categories of expenditure (e.g., central management) that should be imputed to operations and maintenance when tariffs are being determined. Our estimate of current costs in 1988 is $1.24 million for the entire system, or $15.70 per irrigable hectare (Table 6-2). By contrast, INERHI calculated the average volumetric tariff based on only $0.22 million of system-wide operations and maintenance costs (INERHI, 1987).

Measurement of actual water use being rare in rural areas, the amount INERHI charges an individual farmer for current expenses is based on estimated water requirements for his cropping pattern. One of the highest volumetric tariffs is for land planted to potatoes, which require a lot of water. As is indicated in Table 6-2, this upper-limit charge, equal to $7.02 a hectare in 1988, covered less than half the average operation and maintenance costs for INERHI's thirty-five systems.

If anything, INERHI's tariffs, which summed to only 5 percent of the cost of water used to produce potatoes in 1988 (Table 6-2), have exceeded what regional authorities have been charging farmers. In spite of recent increases, CEDEGE and PREDESUR's water prices remain extremely subsidized.

This policy, of course, drives up public sector deficits and contributes to Ecuador's national debt. Multiplying the average per-hectare subsidy, $151.84 (Table 6-2), by irrigable area in all of Ecuador's public irrigation systems (115,000 hectares) yields an annual financial loss of $17.46 million for all projects. As Whitaker, Colyer, and Alzamora (1990, pp. 490 and 504) emphasize, this exceeds the 1988 budgets for INERHI and the National Institute of Agricultural Research (INIAP): $13.17 million and $1.98 million, respectively. If current subsidy rates are not reduced, irrigated agriculture's annual contribution to the fiscal deficit could approach $80 million as new projects are built and put into operation.

Poor Performance of Public Systems

The costs associated with Ecuador's policy of supplying water to farmers at low prices could perhaps be justified if the benefits of public irrigation projects exceeded their costs. Unfortunately, few of those systems meet this criterion.

Of course, poor performance usually contrasts with the original expectations of those who plan and finance irrigation development. For example, the authors of the feasibility study for the Montúfar Project (Figure 6-1) claimed that its internal rate of return would be 26 percent. By contrast, an evaluation carried out five years after it came on line revealed that the actual internal rate of return was closer to 5 percent (IDB, 1981). An ex post facto evaluation of the Milagro Project (Figure 6-1), which is one of INERHI's poorest, also showed a substantial difference between ex ante and actual net returns (IBRD, 1982).

Research carried out by Whitaker and Alzamora (1990a) yields a system-wide perspective on the problem of poor performance. In early 1989, they conducted a survey of real estate values in order to determine the premium offered for irrigated land. (As in other parts of the world, the benefit of receiving cheap irrigation water tends to translate almost entirely into higher real estate values.) Their sample included parcels lying inside systems that account for 61 percent of the irrigable area of the country's government-run projects. Prices for similar parcels close to, but outside, those same systems were also examined. Per-hectare premiums for irrigated land were found to range from $367 to $3,897 and the weighted average was $1,091 (Whitaker and Alzamora, 1990a, p. 183).

This measure of irrigation benefits compares poorly with the subsidies built into most projects. If the average annual subsidy, $152 a hectare (Table 6-2), is sustained for forty years and if the real discount rate is 6 percent, then a per-hectare present value of $2,287 is obtained. In other words, premiums for irrigated land would have to exceed this value in order for a project to be economically viable (assuming that project benefits are not delayed). Any system with a land premium less than $2,287 clearly has a benefit-cost ratio less than one and a negative internal rate of return. Based on the weighted average premium for publicly irrigated land in Ecuador of $1,091 a hectare, farmers receive only about 48 percent of the irrigation subsidy. The balance is completely wasted with no basis for recovery. Society picks up the tab.

Causes of Poor Performance

That agricultural water development often fails to live up to planners' and funders' expectations has much to do with incomplete or tardy project execution. Cost overruns are the rule, not the exception, during the construction phase and investment requirements often turn out to be higher than planned. Also, construction delays and financing difficulties are common. For these interrelated reasons, systems typically begin operating later than planned and many are only partially finished. For example, just sixteen of INERHI's thirty-five operating projects are more than 90 percent complete (Whitaker, Colyer, and Alzamora, 1990, p. 493).

Furthermore, the results of ex ante feasibility analyses often contrast sharply with those of ex post facto evaluations because planning of irrigation projects is sometimes guided by the assumption that every available drop of water will be used for crop production. This would require INERHI to operate its systems throughout the growing season and farmers to irrigate day and night. Either behavior would be a startling departure from current practice.

Particularly inaccurate are feasibility studies' projections of water use and management at the farm level. For example, Keller et al. (1982) found on-farm water efficiencies (evapotranspiration divided by total water delivered to the head of the main farm ditch) to be low in Ecuador: 2 to 15 percent where farm ditches were not lined (which is the typical case) and 20 to 45 percent where ditches were lined. By contrast, international norms are 30 and 50 percent, respectively. Of course, haphazard resource allocation is only to be expected since conservation measures are often expensive and since basic and volumetric tariffs are small.

Largely because of difficulties encountered as irrigation systems have been constructed, there is a major discrepancy between the planned area identified in INERHI's feasibility studies and actual irrigable area once projects have come on line: 140,000 versus 79,000 hectares in 1987. Furthermore, just 52,000 hectares of the irrigable area were actually irrigated that year. The latter shortfall is explained by low efficiencies in water use that, to repeat, are largely explained by subsidized tariffs. (Of course, drought-induced shortages also take a toll from time to time.)

There is another reason why opportunities to take full advantage of dams, canals, and related infrastructure are being lost in Ecuador. For all intents and purposes, there has been very little complementary spending on human capital and agriculture's science base (Chapter 3). After declining for several years, overall investment in agricultural research is very small. In addition, INERHI and other public irrigation agencies dedicate few professional and financial resources to extending information to farmers on the efficient use and management of water. This is only to be expected since their budgets are directed primarily to building and subsidizing irrigation systems. Regardless, inadequate support for technology transfer substantially reduces the benefits of agricultural water development in Ecuador.

Distribution of Benefits

Clearly inefficient, Ecuador's agricultural water systems cannot be justified on social equity grounds. Since economic returns end up being capitalized into agricultural real estate values, the distribution of benefits is a direct function of land ownership, which is highly concentrated in many places. For example, a study of the Pisque system (Figure 6-1) in the early 1980s revealed that 5.5 percent of all the system's farmers owned 45.2 percent of the irrigable area while 52.9 percent of the farmers held just 13.7 percent of the same area (S.A. AGRER N.V./ INERHI, 1982).

Beyond the concentration of benefits among larger land owners within individual systems, equity issues arise at the national level. As pointed out previously, 115,000 hectares—less than 10 percent of all the country's cropland—benefit from projects operated by INERHI and other public agencies. In 1987, expenditures by those same agencies amounted to a little less than 40 percent of all government spending on the sector. By contrast, spending on agricultural research, the benefits of which tend to be more widely distributed, was only 9 percent of the government's budget for the sector during the same year (Colyer, 1990, p. 284).

To summarize, the economic returns on Ecuador's substantial investment in agricultural water development have proven to be poor. In part, this is because the beneficiaries of that development, who are relatively few in number, use and manage water inefficiently. Inefficiency derives largely from the government's policy of selling water far too cheaply. Low returns also reflect inadequate investment in human capital and the science base for agriculture.

Hydroelectricity Development

By no means are waste and misallocation unique to irrigation systems. The benefits of hydroelectric projects in Ecuador do not approach their costs, which are large. A second parallel with irrigation is that energy prices are cheap. A disproportionate share of subsidies is captured by the well-to-do and the burden is shouldered largely by the poor, in the form of inflation brought about because of chronic public deficits.

Major Projects

Hoping to harness its rivers for energy production, Ecuador began to pursue an ambitious plan for hydroelectricity development during the middle 1970s. In some important respects, the achievements have been impressive. Of total installed capacity in the country (1.967 million kilowatts in 1988), almost half is accounted for by hydro plants (BCE, 1991, p. 133). Those plants normally produce 75 to 85 percent of the electricity supply, higher cost thermal plants being used primarily as a backup. Contrasting with the situation a quarter century ago, most of Ecuador now has electric service.

Notwithstanding lip service occasionally paid to small projects, practically all hydroelectricity is generated at a few large facilities, the most important being the Paute complex. Located northeast of Cuenca (Figure 6-2), its first two phases, featuring 500,000 kilowatts of capacity, were completed in 1983 after an investment of $672 million in current dollars, or $872 million at 1988 prices. The third phase of Paute, which added another 575,000 kilowatts to installed capacity, went into operation early in 1992 at a cost of $205 million, or $197 million in 1988 dollars (INECEL, 1988–1992).

Electricity demand has continued to grow, partly because consumption is subsidized. Also, reservoir sedimentation has reduced Paute's effectiveness (see below). So it is that plans to build other dams have continued to be pursued. The Agoyán complex, east of Ambato, was completed in 1987. It added 156,000 kilowatts to installed capacity in Ecuador and cost $125.7 million at 1988 prices. The hydroelectric plant at the Daule-Peripa Dam will require an investment of $150 million and will feature 130,000 kilowatts of generating capacity (Table 6-1). Two other hydro facilities, San Francisco (230,000 kilowatts) and Chespi (167,000 kilowatts), could be constructed before the turn of the century (Figure 6-2).

Inefficient Operations and Theft

In light of the high cost of hydroelectricity development, it is discouraging to note that Ecuador's generating facilities are seriously under-

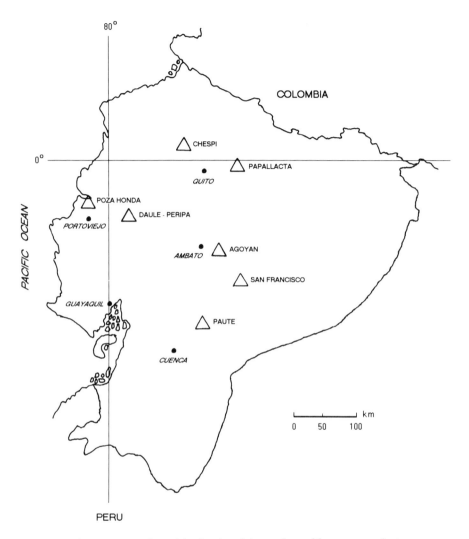

Figure 6-2 Selected hydroelectricity and potable water projects.

utilized. If existing hydro and thermal plants could be run more efficiently, bringing new projects on line would be much less urgent. Additional savings could be captured if transmission losses and theft were reduced.

To put the problem of capacity utilization in perspective, consider that 1.967 million kilowatts of capacity were used to generate 5.603 billion kilowatt hours of energy in 1988 (BCE, 1991, p. 133). This implies that all facilities operated an average of 119 twenty-four days, or one third of the year. By contrast, Pacific Corp operates its thermal and hydroelectric plants in the western United States an average of 292 days a year, or 80 percent of the time (R. Lungman, personal communication, 1991). With an operation 80 percent as efficient as Pacific Corp's, Ecuador could have gotten by with about half its capacity and instead ran its plants 234 days in 1988.

The third phase of the Paute Project seems to be a specific example of excess capacity. During dry months, stream flow is not enough to run the newly installed turbines. When there is enough stream flow, those same turbines are not really needed to satisfy demand inside Ecuador. This latter relationship between consumption and electricity production is tacitly acknowledged by recent INECEL press releases, which indicate that a principal benefit of the latest addition to the Paute complex is to allow the country to sell energy to Colombia and possibly Peru.

Since INECEL has, in effect, invested too much in generating capacity, its financial position is chronically shaky. Financial difficulties are compounded by normal transmission losses, underbilling, and outright theft of electricity. For example, in 1988 paid consumption, 4.372 billion kilowatt hours, was 22 percent less than generation, 5.603 billion kilowatt hours (BCE, 1991, p. 133). By contrast, the gap for Pacific Corp, which results entirely from unavoidable transmission losses, is only 7 to 8 percent (R. Lungman, personal communication, 1991). Finding that losses due just to theft and undercharging range from 4 to 22 percent for seventeen of Ecuador's local electric utilities, Soria (1989) estimated that each percentage point of such losses costs the public utility system $1.8 million a year in terms of lost revenues.

Price Subsidies

Waste and misallocation do not result exclusively from inefficient operation of hydro and thermal plants, high transmission losses, and theft. Setting prices well below generation and delivery costs has caused consumption to be excessive and has exacerbated the financial difficulties of INECEL and local utilities. In addition, excessive demand growth has obliged the country to make needless additions to generating capacity.

The magnitude of electricity subsidies can be inferred from a comparison of Paute's capital costs and average energy prices charged in Ecuador. Our estimate of capital costs, $65.04 million annually or $0.03 per kilowatt hour, reflects 2.8 billion kilowatt hours of energy generation at Paute in 1988 (INECEL, 1988–1992). In addition, we assume that 22 percent of that production is lost (see above), that the total investment of $872 million is to be amortized over the complex's working lifetime, which has been estimated to be twenty-eight years (Southgate and Macke, 1989), and that investment spending carries a real discount rate of 6 percent. Significantly, electricity prices in Ecuador averaged $0.02 per kilowatt hour for all classes of users in 1988 (INECEL, 1992), which is just two thirds of Paute's capital costs. Insofar as current costs are not covered, the return on Ecuador's largest hydroelectric investment is sharply negative.

The financial burden of low electricity prices is absorbed largely by INECEL. For example, its operating deficit in 1988 was $51 million, with expenditures totaling $101 million and revenues from all sources amounting to just $50 million (BCE, 1989, pp. 199–200). However, those revenues included earmarked petroleum funds of $12 million as well as $18 million of other governmental transfers. Customers' payments were only $20 million, which was a little less than a fifth of the institute's outlays. Those outlays were reduced because INECEL did not have to pay world market prices for fossil fuels. Its consumption of bunker fuel, used in thermal plants, and diesel was 28,089,000 gallons and 1,241,000 gallons, respectively (INECEL, 1988). Domestic prices, $0.160/gallon for bunker and $0.298/gallon for low-grade diesel, were less than half the values in international markets in 1988 (BCE, 1990). Without subsidies, INECEL's fuel expenditures and its overall losses would have been about $5 million higher.

Distributional Consequences

As is the case with irrigation subsidies, a disproportionate share of the benefits of cheap electricity are captured by wealthy households. Kublank and Mora (1987) have found that the richest 10 percent of the residents of Quito and Guayaquil consume 90 percent of the fossil fuels and electricity sold in those two cities. But in spite of its regressivity, Ecuador's cheap energy policy receives virtually unanimous support. As opposition politicians are quick to point out, the poor suffer (in the form of higher costs for food and other goods and services) whenever government-controlled fuel or electricity prices are increased. So it is that the government has been slow to raise tariffs enough to eliminate INECEL's deficits (IBRD, 1991a, pp. 108–110).

Those who think the interests of Ecuador's poor are best protected by subsidizing energy consumption ignore the macroeconomic consequences of that policy. At the end of every fiscal year, the central government extends credits to INECEL in order to cover the latter's deficits. For example, internal credits of $46 million were used to cover the 1988 reported operating deficit of $51 million (see above). (INECEL also took in $5 million in external credits.) In other words, public utility deficits were monetized. The poor, needless to say, were not shielded in any meaningful way from the resulting inflation.

Potable Water Systems

The economic record of Ecuador's potable water projects is no different from that of hydroelectricity and irrigation development. A substantial investment has been made to meet rapidly expanding demand, and consumption of potable water is subsidized. The poor have usually been the last to benefit.

Recent History of Potable Water Development

Providing all Ecuadorians with clean water, which is essential for containing the spread of infectious diseases, is a major challenge. Supplying the needs of rural households is often costly. In addition, municipal water and sewer systems have been severely strained as urban populations have burgeoned. During the 1980s, many parts of Quito suffered chronic shortages. Keeping Guayaquil adequately supplied has also been difficult, especially during the coastal dry season when seawater can seep into the municipal system due to low stream flow in the Daule River. In some parts of the country, agriculture has felt the consequences of increased demand for potable water. For example, the share of water released to irrigation systems from Poza Honda Reservoir (Figure 6-2) has declined over the years due to the rapid growth of Portoviejo and other urban areas in Manabí province.

Small-scale initiatives to improve potable water supplies in the countryside have been supported by the national government, international donor agencies, and private organizations. In addition, major projects, financed with oil revenues and donor funds, have been undertaken to alleviate water shortages in Ecuador's cities and towns. Inaugurated in 1990, the Papallacta system benefits Quito. A major impact of the first phase of the Daule-Peripa Project has been to make stream flow in the lower Daule River less variable (Table 6-1). In particular, average flow during the last half of the calendar year, when rainfall ebbs, has increased from 55 to 100 cubic meters a second (Arriaga, 1989, p. 151), which has helped to free Guayaquil's municipal system from saltwater intrusion.

Subsidized Prices

Since the Daule-Peripa Project serves several purposes (flood control and navigation in addition to providing water to urban areas and farmers), determining the cost of providing for any particular group of users can be difficult. By contrast, estimation of the subsidies enjoyed by Papallacta's beneficiaries is relatively straightforward. Although turbines could be installed there, the reservoir is used exclusively as a source of potable water for Quito, with withdrawals averaging 3 cubic meters a second.

Capital costs for the project, which took three years to construct, beginning in 1988, were $122.3 million in current prices and $130.9 million in 1991 U.S. dollars (Município de Quito, 1992). If a thirty-year project life and a 6 percent real discount rate are assumed, then amortization amounts to $9.5 million a year. Adding in expenditures for operations, maintenance, and periodic replacement of equipment, which were $6.2 million in 1991 (Município de Quito, 1992), and assuming that water is delivered twenty-four hours a day all year, the average cost of Papallacta water in 1991 was $0.166 per cubic meter. That average cost is nearly triple the 1991 price of $0.061 a cubic meter (at the 1991 average exchange rate of 1,103.43 sucres to the dollar) for a Quito household consuming 30 cubic meters a month (EMAP-Q, 1991).

The preceding analysis of costs and prices understates total subsidies. For one thing, cash outlays for the municipal water authority are reduced by fuel subsidies. Although it is not documented, some underbilling of water occurs in Quito. In addition, spending on additional sewer capacity, which is needed to accommodate increased outflow from the city, has not been taken into account.

Distributional Consequences

Municipal governments generally try to make potable water systems comprehensive. But some neighborhoods end up being neglected. For example, it is nearly impossible to extend pipes throughout the "Guasmo," Guayaquil's principal slum (where hundreds of thousands of people live), since much of it is built over swamps. Similarly, some of the marginal neighborhoods that have sprung up in the hills around Quito are not served by the municipal system.

A large number of poor households depend exclusively on water delivered by tank trucks. Prices are an order of magnitude greater than municipal tariffs. For example, $1.50 was being paid for a cubic meter of tank water in Quito's poor districts in April 1991, when the cost of Papallacta water was only $0.17 a cubic meter (see above). Although price gouging is frequently alleged, competitive market forces are at work. In other words, charges for tank water are not far out of line with the cost of providing the service.

It is fair to say that the poor tend to be served least well by municipal systems. While waiting for pipes to be extended out to their neighborhoods, they bear not only high prices for tank water but also the inflation tax arising as central government credits are provided to cover local authorities' operating deficits.

Water Resource Degradation

Often quick to spend money on new projects and to subsidize consumers, the Ecuadorian government has failed to manage watersheds and to reduce pollution. This omission has much to do with the budgetary demands of traditional patterns of water resource development, as described in this chapter. If nothing is done to halt watershed deterioration and to improve the quality of lakes, rivers, and streams, though, future development could be hindered.

Deteriorating Watersheds and Reservoir Sedimentation

Agricultural frontiers have been expanding throughout Ecuador, in tropical moist forests as well as the Sierra. Especially in Andean hillsides and *páramos*, crop and livestock production can lead to accelerated soil erosion (Chapter 5). The downstream consequences of this process include lost reservoir capacity, increased flooding, and diminished water harvesting.

Most of Ecuador's hydroelectric dams lie at the base of highland watersheds (Figure 6-2), the upper reaches of which tend to be steeply sloped and covered with erodible soils. The consequences of failing to take into account the sedimentation that is likely to result because of these natural conditions and also from agricultural frontier expansion in the highlands are illustrated by what has taken place at the Paute complex in recent years.

The country's major generating facility has been plagued by design flaws. For example, the discharge tunnel at the bottom of the dam, which is used to flush sediments out of the reservoir and thus keep intakes to the turbine complex clear, is too small. It is close to being covered entirely.

These problems are indicative of a general failure to anticipate the magnitude of sedimentation at Paute. The project's lifetime will probably be considerably less than fifty years, which was assumed in feasibility studies. Assuming that dredging close to the dam, to prevent clogging of turbine intakes, would begin six years after the complex came on line in 1983, Southgate and Macke (1989) estimated that Paute would have to be closed after operating for twenty-eight or twenty-nine years. Even this finding might be optimistic. Dredging did not actually get under way until 1991 and watershed management activities, for which INECEL has received $14.5 million from the Inter-American Development Bank (IDB), have barely started. In early 1991 and again in early 1992, cumulative sedimentation and below normal precipitation combined to force a reduction in Paute's energy output. Ecuador suffered economically debilitating blackouts and power rationing as a result.

To its credit, INECEL seems to have learned from past mistakes. Designed to hold water for long periods of time, the Paute Reservoir originally had 120 million cubic meters of storage capacity. Although hard numbers are difficult to come by, much of that volume has been displaced by rocks, soil, and other eroded material. By contrast, Agoyán, which opened four years after Paute did, is a large run-of-the-river complex, one in which stream flow, rather than water released from a large reservoir, drives the turbines. Sediments can be evacuated out of the reservoir, which holds a little more than 2 million cubic meters of water, as often as twice a week. Although it has very little capacity to store water until it is needed at times of low rainfall, Agoyán has turned out to be a reliable source of hydroelectricity during recent dry spells. (Of course, frequent reservoir flushing can impose costs on downstream riparian land users.)

Flooding and Reduced Water Harvesting

Even though Ecuador has always suffered from seasonal flooding, especially in the Costa, it is frequently claimed that episodes of high water have become more serious in the last few years. One explanation is probably increased human settlement of floodplains. However, upper watersheds have lost forests and other natural vegetation. As a result, water runoff during and immediately after major storms has become more pronounced.

Rather than controlling settlement close to rivers and streams and conserving tree cover and soils in upper watersheds, the Ecuadorian government tends to adopt structural solutions to flooding, often with the aid of international donor agencies. For example, an important use of the Daule-Peripa Dam is to regulate stream flow at lower elevations. With funding from the World Bank, CEDEGE also intends to undertake complementary flood control projects in the lower Guayas River Basin.

The effects of rapid water runoff, soil erosion, and sedimentation and flooding in Ecuador are well understood, though not very well quantified. Practically nothing is known about the effect on water harvesting of watershed destruction. Empirical evidence is limited, and the concept of water harvesting is novel in the country, even to some government policymakers and professional and technical personnel involved in water projects. What is certain is that the deterioration of upper watersheds leads to reduced capacity to store water in the soil and less available water from a given amount of precipitation. Rainfall quickly runs off eroded slopes and down streams and rivers and is discharged into the Pacific Ocean or carried beyond Ecuador's eastern borders. With proper watershed management, more water could be retained for use in agriculture, industry, and private homes. This benefit would be particularly important in those parts of the Sierra with little rainfall.

Water Pollution

Pollution has grown severe in many parts of Ecuador. Emissions of untreated sewage and industrial wastes from cities and towns have contaminated a number of waterways. For example, pollution from Guayaquil is a principal cause of high bacterial contamination, low dissolved oxygen demand, and high nutrient concentrations in the Guayas River and its tributaries (Solórzano, 1989). Similarly, contamination of the Machángara River, which flows through Quito, is so severe that using it as a source of irrigation water is dangerous (Landázuri and Jijón, 1988). Regardless, a few people still use river water to wash clothes in and even to drink.

Urban activity is not the only threat to water quality and, by implication, to human health. Kimerling (1991a, 1991b) stresses that oil exploration and development in the Oriente has been accompanied by pronounced environmental degradation and water pollution (Chapter 7). In addition to suffering the impacts of negative externalities produced by other sectors of the economy, agriculture is a source of pollution. Agricultural water pollution takes the form of runoff of fertilizers, pesticides, and other chemicals. In addition, water quality problems arise in coastal reservoirs, which have been constructed for irrigation development and other purposes.

Public concern over the misuse of agricultural chemicals was first aroused in 1985, when Fundación Natura (Ecuador's leading environmental organization) issued its first study of pesticide contamination. Particularly alarming was the report that twenty-three substances that most affluent countries either ban entirely or carefully restrict the use of were being applied on farms in Ecuador (Sevilla-Larrea and Pérez de Sevilla, 1985), presumably because bans and restrictions have made the same substances relatively cheap. Subsequent research showed that hazardous pesticides had found their way into food eaten by people. Human breast milk in Esmeraldas, Guayaquil, and Quito, the only three cities studied, was found to contain traces of benzene hexachloride (BHC), aldrin, DDT, and other banned substances (MAG/CONACYT, 1986).

As in many other nations, analysis of the movement of agricultural chemicals through waterways, up the food chain, and into human bodies remains at an incipient stage in Ecuador. With funding from the Rockefeller Foundation, scientists from North American universities, the International Potato Center (CIP), and Ecuador have been studying the impacts on human health and water quality of heavy pesticide use by the potato farmers of Carchi province (Crissman and Espinosa, 1992; Crissman and Cole, 1994).

Documentation of high concentrations of fertilizers and chemicals in other parts of the country suggests that this kind of research needs to be extended. For example, Solórzano (1989) has found that waterways in the lower Guayas River Basin, where agriculture is more highly developed than anywhere else in Ecuador, contain substantial amounts of nitrates and pesticides. In the same region, run-off of pesticides, which are being used to contain the spread of Black Sigatoka (a fungal disease) in banana plantations, appears to be reducing production in the shrimp industry, which is one of the leading generators of foreign exchange in Ecuador (Chapter 8).

Just as extreme sedimentation problems have arisen in the Paute Reservoir, in the southern highlands, water quality has proven difficult to maintain in large impoundments at lower elevations. Dissolved oxygen concentrations have been chronically low in the Poza Honda Reservoir. In addition, water hyacinths (*Eichhornia crassipes*) have proliferated in the lake created by the Daule-Peripa Dam.

Water hyacinth infestation can make life difficult for boat operators in the reservoir. It could also interfere with energy production, as contemplated in the second phase of the project (Table 6-1). Specifically, plants could clog intakes and perhaps even damage turbines.

There is no easy way to get rid of aquatic plants. Removing water hyacinths mechanically is expensive. As an alternative, CEDEGE sprayed 2,4-d (dichlorophenoxyacetic acid) in the Daule-Peripa Reservoir in 1990 and has considered a repeat application. However, using enough of the herbicide to kill off all the weeds could contaminate water supplies for downstream populations. At an extreme, the water Guayaquil normally withdraws from the Daule River could be rendered unfit to drink.

One possible solution with virtually no adverse environmental effects is to introduce manatees or some other large herbivore. Even engineers, who tend to look first for a mechanical or chemical approach, are starting to give this option serious thought.

Summary and Conclusions

For the most part, environmental problems analyzed in this book arise because of excessive governmental interference with market forces and private property rights. As is documented in this chapter, waste and misallocation of Ecuador's water resources is an excellent case in point. Since nationalizing those resources twenty years ago, the public sector has spent hundreds of millions of dollars on dams, canals, and related infrastructure. The return on this investment has been extremely poor, both because consumption of water and hydroelectricity has been heavily subsidized and because, in the case of irrigation systems, complementary spending on education, research, and extension has been inadequate.

An irony of the recent approach to water resource development is that delivery of services has actually gotten worse for many segments of the population. Secondary irrigation canals that used to function quite well have fallen into disrepair since the Water Law nationalized water resources. Also, failure to plan adequately for reservoir sedimentation and periodic shortfalls in precipitation has meant that electricity supplies have grown less realiable over time for many customers.

If policies are not changed, the problems reviewed in this chapter will grow worse. Due to subsidies, pressure to pursue new projects, most of which can be expected to be inefficient and to have regressive distributional consequences, is constant and strong. The deficits of agencies involved in water resource development are monetized. As a result, the poor, who tend to be the last served by irrigation, potable water, and electricity systems, pay a heavy inflation tax.

One other consequence of inefficient investment in water projects is indicated at the end of this chapter. Since INERHI, INECEL, and local and regional authorities lose tens of millions of dollars a year, the government is left without the means to protect the quality of its aquatic "properties." As we have pointed out, Ecuador is doing almost nothing to manage watersheds and to reduce water pollution. Though not quantified, the costs of this neglect, in terms of lost opportunities for water resource development and, worse yet, the high incidence of water-borne diseases and parasites, are very high in the country.

7

Oil Industry Pollution in the Ecuadorian Amazon

Since 1967, when major petroleum deposits were discovered in the Oriente, the Ecuadorian economy has come to depend on fossil fuel production. The oil industry generates about 15 percent of the country's GDP and nearly half of its exports. In addition, petroleum revenues account for 40 percent of all public sector income. To keep the national economy afloat, over a thousand wells have been drilled and an accompanying network of roads, pipelines, and related facilities has been established.

Environmental quality has been sacrificed for the sake of accumulating oil wealth. Formation water, which is extracted along with petroleum, is routinely discharged into rivers, lakes, and streams. Large quantities of oil have also found their way into the environment, in part because of damage caused to pipelines by earthquakes and other natural disasters. In addition, faulty construction and maintenance of storage pits and pipelines is a cause of water pollution, which has been linked to various health problems.

In this chapter, the environmental impacts of petroleum development are examined. Evidence concerning the magnitude of oil industry pollution is summarized, as are national laws and regulations intended to maintain water quality. We also address the controversy over expanded operations in the Oriente, which has reached international proportions, through an economic analysis of Block 16, which is one of the new concessions.

Environmental Impacts

In a pair of recent publications, Kimerling (1991a, 1991b) argues that tropical forests and their inhabitants start to suffer once initial contact is made with the oil industry. Exploration begins with seismic studies that last one to two years and that are accompanied by noise, which disturbs game, and by limited forest disturbance. Subsequently, wells are drilled to carry out more detailed resource evaluations. Around each exploration platform, 2 to 5 hectares are completely deforested and construction timber is taken from another 15 hectares or so.

Approximately 1,000 barrels of oil are extracted from a typical exploration well. Most of this oil is burned. In addition, drilling muds, which contain toxic substances in low concentrations, and mud-coated drilling wastes are dumped into reserve pits, many of which are uncovered and prone to seepage. On average, each well produces nearly 4,200 cubic meters of muds and wastes during its lifetime (Kimerling, 1991b, p. 862).

Much of the damage accompanying exploration is confined to a small area and of limited duration. Environmental impacts become much more serious once production begins. On average, 325 barrels of liquid wastes are discharged each day from the Oriente's production wells, of which there are almost 1,100. In addition, carelessness and inadequate maintenance lead to losses of oil from the flow lines connecting production wells with the facilities where petroleum is separated from natural gas and formation water. These losses are reported to be between 29 and 36 barrels a day (Kimerling, 1991b, p. 865).

Since natural gas concentrations in the Oriente tend to be fairly low, flaring off is common practice. There is only one gas plant in eastern Ecuador, at Shushufindi. It uses no more than 15 percent of what is pumped out of the ground in the region. Burning the remainder is not too much of a problem since air quality is generally very good.

By contrast, separation facilities are a major source of water pollution. At those sites, more than 100,000 barrels of untreated production water, comprising chemicals needed to break the bonds between oil and formation water as well as formation water itself (which has an average temperature of 54°C when it comes out of the ground, is alkaline, and contains heavy metals), are discharged each day into unlined production pits. Virtually all the production water held in the pits eventually finds its way into the environment.

After the separation facilities, oil (with a water content of less than 0.1 percent) is channeled into the Trans-Ecuadorian Pipeline, which runs over the Andes to the port of Esmeraldas (Figure 7-1). The national government has recorded thirty major spills from the pipeline in the first twenty years of operations, the most spectacular one occurring during the earthquake of March 1987. Kimerling (1991b, p. 872) calculates that

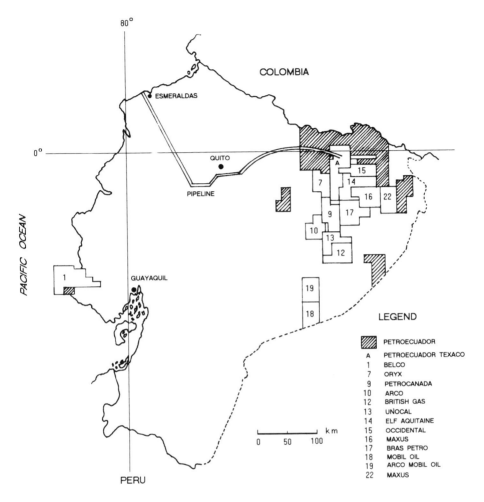

Figure 7-1 Oil concessions in Ecuador.

400,000 barrels have been lost in those thirty events. This figure seems a little high since no more than 50,000 barrels were spilled when 25 kilometers of the pipeline were destroyed in the 1987 earthquake.

For the first twenty years after oil was discovered near Nueva Loja, there was very little monitoring of the environmental impacts of petroleum development. The first major study of surface water quality in the Oriente was released in 1987. Thirty-six sites, all located close to exploration or production facilities, were sampled. Petroleum concentrations were found to be high everywhere and, at most places, dissolved oxygen was below the levels needed to maintain healthy aquatic communities (CEPE, 1987).

No systematic investigation of the social impacts of pollution has been conducted. However, anecdotal evidence is alarming. People living near oil wells complain of diminished fish catches, skin rashes after bathing in streams, and nausea after eating contaminated food or drinking polluted water. In addition, local communities suffer when employees of oil companies or their contractors trample garden plots or fish in a depletive fashion, to give two examples of sloppiness and insensitivity (Kimerling, 1991a).

Current Regulatory Environment

As is the case in many other Latin American nations, the Ecuadorian constitution guarantees each of its citizens the right to an "environment free of contamination." In the service contracts that Petroecuador, the state oil company, signs with private firms operating in the country, there are admonitions to protect flora and fauna and to prevent air, water, and land pollution. In addition, the Hydrocarbons Law of 1990 calls on Petroecuador and its contractors to develop "plans, programs, and projects" to conserve renewable resources and to benefit local communities. But as conditions in the Oriente demonstrate, the specific legal guidelines and enforcement mechanisms needed to effect these general standards have not been developed.

Before 1984, when the Ministry of Energy and Mines (MEM) set up a General Directorate of the Environment (DIGEMA), no government agency had a direct mandate to hold the oil industry to any environmental standard. From the beginning, DIGEMA was starved of political and financial support. In 1988, it launched its most ambitious initiative, informing oil companies that environmental impact statements had to be submitted and approved before any new exploration or production project could begin. For the most part, this requirement was ignored. Of more lasting significance was DIGEMA's preparation of guidelines for exploration and development in protected areas, which have served as a point of departure for the subsequent drafting of regulations to be applied throughout the country.

In 1990, DIGEMA was renamed the National Directorate of the Environment (DINAMA) and incorporated into MEM's new Subsecretariat of the Environment (SMA). Soon after the reorganization, DINAMA submitted draft environmental regulations for the oil industry, which had been two years in the making. However, those regulations were never adopted. Instead, a "gentleman's agreement" was signed by SMA, the oil companies, and Fundación Natura. The agreement called on the companies to observe a loosely structured set of environmental guidelines for two years, although it was largely silent on the issue of enforcement. At the end of that period, SMA was supposed to come up with detailed regulations that had to be accepted by all signatories.

There continue to be ample grounds for skepticism about the Ecuadorian government's interest in curbing oil industry pollution. In early 1991, DINAMA had only four full-time technical staff members, all based in Quito, even though it retained a mandate to monitor petroleum development in all parts of the country. Around that time, the head of SMA levied three maximum fines, which amounted to only $2,000, on Petroecuador and Texaco (the state company's principal contractor) for severe pollution in the Cuyabeno Wildlife Reserve, and the Limoncocha Biological Reserve in northeastern Ecuador. He was promptly denounced for causing MEM and Petroecuador not "to speak with a single voice." In March, he was sacked.

The government's true commitment to curbing oil industry pollution remains in doubt. In February 1993, the state oil company established three exploration camps in the Cuyabeno Reserve, without prior authorization from INEFAN, which manages all protected areas in the country. After a widely viewed television report in late April, Petroecuador submitted a "diagnostic study," which contained a series of recommended actions, on May 17. It filed an environmental management plan, as a complement to the study, after a meeting that INEFAN convened on June 9. At that meeting, which was attended by representatives of local indigenous communities and the ecotourism industry, one maxium fine, of $315, was levied on Petroecuador. Of much greater importance was INEFAN's decision to suspend all exploratory work in the reserve, which was upheld by the president of the country in July. However, Petroecuador, which had already signed contracts with various suppliers and outfitters for the Cuyabeno project, succeeded in convincing the president that the suspension created an excessive financial burden. Around the middle of September, INEFAN's director was informed that drilling would go forward immediately at one of the three contested sites.

The International Controversy Surrounding Development of Block 16

Pollution and related environmental problems in the Oriente, and the lack of an effective regulatory response by the government, have generated strong opposition to expanded petroleum development in the region. Contamination and unauthorized drilling in the Cuyabeno Reserve and elsewhere have led to domestic scrutiny of Petroecuador's operations. In addition, companies based in North America and Europe have been pressured to abandon their Ecuadorian projects.

As Southgate (1992b) has reported, the international debate has recently come to a head in one single concession: Block 16 (Figure 7-1). Part of the concession lies in Yasuní National Park, one of the world's most species-rich rainforests. It also extends into the reserve assigned to the Huaorani, one of the last indigenous groups living in relatively isolated conditions in the Oriente.

The Campaign against CONOCO

Having discovered commercial deposits in Block 16, CONOCO, a Du Pont subsidiary, worked on development for several years. Pressed by environmentalists in Ecuador and around the world, it attempted to control pollution and to limit adverse impacts on forest dwellers. During the exploration phase, the company contained the spread of disease among the Huaorani and other tribes by inoculating them and also by limiting outsiders' access to the concession. Further, the number of production sites was to be minimized (by drilling directionally from a few sites instead of vertically from many) and formation water was to be injected back into the ground. Finally, CONOCO was not going to build a bridge across the Napo River, which is a natural impediment to colonization in and around Block 16.

Although these measures placed CONOCO among the more environmentally responsible foreign oil firms operating in Ecuador, some defenders of Ecuador's tropical forests felt obliged to apply strong pressure on CONOCO to abandon its operations in the Oriente. In 1990, the Sierra Club Legal Defense Fund petitioned the Inter-American Commission on Human Rights on behalf of the Huaorani to stop development of Block 16. That petition having been rejected, a group of environmental lawyers from Oregon tried to employ the Foreign Corrupt Practices Act for the same purpose. In June 1991, the latter attempt was rejected by the U.S. Justice Department.

Lawsuits having proven futile, environmentalists mounted a lobbying campaign. The goal of this campaign was stated clearly by Randy Hayes, of the Rainforest Action Network (RAN), who promised, "There's one thing you can be sure that we at RAN will never do, and that's compromise with CONOCO."

RAN and its allies no longer have the opportunity to do so. On October 11, 1991, CONOCO announced that it was quitting Ecuador. Maxus, a smaller U.S. firm, is the new operator of the Block 16 concession.

The Campaign's Impacts

The importance of lobbying by RAN and its allies should not be exaggerated. An important cause of CONOCO's withdrawal was the grindingly slow pace at which the Ecuadorian government negotiates concession agreements. The company had to wait two years before its proposal for development of Block 16 was approved by Petroecuador. When it announced that it was giving up its concession rights, CONOCO was still waiting for MEM approvals.

Other firms have been driven out of the country for the same reason. Petrocanada expressed an interest in investing $400 million in the offshore gas fields of the Gulf of Guayaquil. Ecuador's first fertilizer plant was to be constructed as a result. Frustrated by the government's inability to make decisions, the company did not pursue the project.

Another factor discouraging oil industry investment in Ecuador is the presence of attractive options elsewhere. CONOCO might have determined that bringing new deposits on line in the country is not very profitable (see below). When it pulled out of Block 16, it was looking to expand into Russia and the North Sea.

The true significance of the campaign by RAN and its allies is that the difficult situation faced by environmentally responsible elements of Ecuador's petroleum industry is being made worse. All companies operating in the country do so under a service contract with Petroecuador, which must approve all expenditures. Unfortunately, neither the state company nor the national government has developed a detailed set of environmental guidelines for petroleum exploration and extraction. Also, some contractors complain that Petroecuador is slow to approve environmentally related spending.

Under these circumstances, foreign companies find it difficult to convince skeptics that their Ecuadorian operations will indeed be conducted in an environmentally sound manner. When those skeptics decide to target a company (as has been the case with CONOCO, Atlantic Richfield Company, and other U.S. firms and with British Gas in the United Kingdom), the company is strongly tempted to leave Ecuador, particularly if it feels obliged to try to please the green lobby.

The Economics of Pollution Control

Unfortunately, national and international debates over Block 16 and other petroleum operations in eastern Ecuador are being carried out with little or no regard for economic trade-offs facing the oil industry and the national government. Controlling pollution, after all, costs money. Due to higher expenditures, private firms might not want to proceed with development. In addition, the government might find that tax revenues must be sacrificed in order to improve environmental quality.

These possibilities are addressed in the remainder of this chapter. To investigate the chances that investment will be discouraged, the internal rate of return associated with development of Block 16 is estimated for two cases: with and without pollution control. In addition, the linkage between environmental standards and the present value of income taxes

86   Case Studies

and production royalties generated by the same concession is explored. Some of the results of this analysis were originally presented in Southgate (1992b).

Data and Assumptions

As is indicated in the preceding section, several measures have been used in Block 16 to limit environmental damage. The number of production sites will be reduced by drilling directionally from a few locations, instead of vertically from many. In addition, formation water will be reinjected and synthetic "geogrid" material, instead of logs cut from the surrounding forest, has been used as a road base.

Directional drilling and reinjection could enhance oil production. Unfortunately, data needed to examine this benefit are not available. In this study, only the impacts on preproduction expenses and operation and maintenance costs are considered. Environmental measures are expected to account for roughly 10 percent of the $600 million that will be spent to bring Block 16 on line and for approximately 15 percent of operation and maintenance costs, which will be about $3 a barrel.

After a run-in period of five years, normal production (45,000 to 50,000 barrels a day) will begin in 1999 (Southgate, 1992a). That output level is expected to be maintained for about thirteen years, after which there will be a gradual decline.

No attempt is made here to assess the specific terms of the agreement signed by Petroecuador and the service contractor working on Block 16. Instead, the site's overall profitability is examined. It can be taken for granted that, if the internal rate of return is greater than or equal to a normal level of profitability in the oil industry, contract terms that will elicit the investment needed to develop the oil deposit can be found.

In addition to depending on investment and operation and maintenance costs and production levels, Block 16's overall profitability depends on loan interest rates, taxes, and oil prices. A real interest rate of 10 percent is assumed in this analysis. Payments to the government include a production royalty, amounting to 3 to 4 percent of the gross value of output, as well as a 42.5 percent tax on income (Southgate, 1992a, 1992b). Analysis has been done for three average oil prices. The base value is $15 a barrel, which is close to what Block 16's moderately heavy and highly viscous oil will probably sell for in the coming years. Analysis is also done for lower and higher values: $10 and $20 a barrel.

Results

Reported in Table 7-1 are estimates of Block 16's earnings under various assumptions regarding oil prices, taxes, and pollution control expenditures. Nothing influences the internal rate of return more than output

Table 7-1 Internal Rate of Return on Block 16 Investment (%)

Assumptions	Oil price		
	$10/barrel	$15/barrel	$20/barrel
No pollution control and current tax rates	9.5	20.9	29.3
Pollution control and current tax rates	6.5	18.2	26.6
Pollution control and 50 percent tax rate cut	9.2	22.4	32.2

value. Raising the price from $15 to $20 barrel, for example, causes profitability to increase by 40 to 55 percent. By contrast, improved pollution control diminishes the internal rate of return by only a little more than an eighth when oil sells for $15 a barrel (Table 7-1).

Industry sources say that private investment would not occur if the internal rate of return were much less than 20 percent. Accordingly, it is likely that development will proceed with pollution controls under existing tax arrangements if oil prices stay around $20 a barrel. At the intermediate value, the project just meets the minimal profitability criterion, but only if there are no pollution controls. At lower prices, Block 16 is a poor investment, even if no effort is made to contain environmental damage (Table 7-1).

Earnings could, of course, be raised by lowering taxes. For example, if real oil prices stay around $15 a barrel, then Block 16 can be developed in an environmentally sound fashion if income tax rates are halved (Table 7-1). However, its internal rate of return is only 16 percent if oil prices stay at $10 a barrel, even if there are no pollution controls, production royalties, and income taxes.

Decisions about environmental standards to be applied in the petroleum sector are not driven exclusively by a desire to attract private investment. The Ecuadorian government, like the citizens it represents, also wants to reap production royalties and income taxes.

Reported in Table 7-2 are estimates of the present value of these two returns, which are based on the assumption of a 10 percent real interest rate. Those estimates indicate that, as long as tax rates do not change, the relationship between government revenues and environmental expenditures is weak. If oil sells for $10 a barrel, adding $60 million to preproduction spending and $0.45 to operation and maintenance costs diminishes the present value of royalties and taxes by 15 percent. At higher oil values, the relative impacts of pollution control on tax payments are smaller.

But if taxes have to be cut in order to bring Block 16's internal rate of return up to minimal industry standards of profitability, then there will

Table 7-2 Present Value of Income Taxes and Production Royalties Associated with Block 16 Development ($ in millions)

Assumptions	Oil price		
	$10/barrel	$15/barrel	$20/barrel
No pollution control and current tax rates	169	367	574
Pollution control and current tax rates	145	333	540
Pollution control and 50 percent tax rate cut	89	191	303

be a large sacrifice in government revenues. For example, halving income taxes would cause the present value of those revenues to fall by 48 percent under the case where oil sells for $15 a barrel and pollution controls are applied (Table 7-2).

Discussion

The preceding analysis is somewhat biased against pollution control. The additional production associated with reinjecting formation water and drilling at an angle, which also benefit the environment, has been ignored. In addition, some knowledgeable observers suggest that industry estimates of pollution control costs, which are reported in this chapter, might be excessive.

In spite of this bias and data limitations, our findings yield clear insights into the economic trade-offs associated with developing petroleum deposits in an environmentally sound fashion. If the medium-quality crude oil that Ecuador has yet to exploit can be sold for $20 a barrel, then investors can earn normal profits, the environment can be protected, and the government can collect sizable production royalties and income taxes. By contrast, normal profits cannot be earned if output is worth only $10 a barrel, even if there are no pollution controls and taxes. At that price, Ecuador would be better off to leave resources untapped until some future date, when their value will presumably be higher.

At the intermediate price used in this study, which is fairly close to long-term average values, the country faces some difficult policy choices. One option is to develop resources without pollution controls so that the public sector can collect more revenues. Alternatively, production can go forward with less damage to the environment, but only if those revenues are reduced.

Summary and Conclusions

Two types of mining are taking place in Ecuador's tropical forests. The first, petroleum extraction, is obvious to any student of the country's economy. Referring to Schneider et al.'s (1991) analysis of the behavior of loggers, colonists, and other agents of deforestation in the Brazilian Amazon, one can identify a second type of mining, which is the depletion of forest ecosystems.

The populations engaged in each type of resource extraction are almost entirely distinct. A recent World Bank report on the Oriente stresses that the Ecuadorian oil sector is a classic example of an enclave industry, one with few backward or forward employment linkages (Hicks et al., 1990). Its relatively small work force is largely segregated from surrounding communities, the economic life of which revolves around ecosystem mining.

This is not to say, of course, that linkages between petroleum extraction and ecosystem mining are weak. Road construction, undertaken primarily to facilitate oil development, has been a necessary condition for agricultural colonization in northeastern Ecuador (Chapter 4). In addition, local residents are susceptible to water pollution and other negative externalities associated with petroleum exploration and production (Kimerling, 1991a, 1991b).

Oil development could make a greater contribution to reducing the dependence of the majority of the Oriente's population on renewable resource depletion. For example, more oil wealth could be channeled into education as well as research and extension for agriculture, forestry, and other sectors of the rural economy. Formation of nonenvironmental assets would, in turn, facilitate sustainable economic development.

Unfortunately, investment in human capital and the science base in the Oriente has been seriously deficient. There is a strong factual basis for the widely circulated complaint that a small portion of oil earnings is made available for education and health services in the region (Hicks et al., 1990). Agricultural research and extension services are practically unavailable. In effect, Amazon rainforests and their inhabitants have been mined for the good of the rest of the country.

Ecuador does not have a great deal of time to reverse this situation. As of 1988, proven reserves were roughly equivalent to cumulative oil production in the Oriente (Hicks et al., 1990). If patterns of development are not changed soon, the region's rivers will be polluted and its land depleted. What is left of its human population will be devastated.

8

Shrimp Mariculture and Coastal Ecosystems

Since the late 1970s, Ecuador has become the Western Hemisphere's leading producer and exporter of shrimp. From modest beginnings, the country's shrimp industry has grown to the point where it often generates more foreign exchange than any other part of the rural economy.

Increased production has come about largely through maricultural development. About 8,000 metric tons have been captured off the Ecuadorian coast each year since the middle 1970s. Meanwhile, pond output has increased severalfold, from less than 5,000 metric tons in 1979 to more than 100,000 metric tons in the early 1990s.

In its earliest years, Ecuadorian mariculture was exclusively "extensive," with the sides of ponds located in low-lying areas occasionally being opened to let in seawater containing shrimp postlarvae (PL) and nutrients. With time, more operations have adopted "semiextensive" production technology, which is characterized by higher PL stocking and mechanized exchange of seawater for pond water (which is needed for rudimentary control of water quality). As the industry continues to evolve, some enterprises are becoming "semi-intensive," employing more sophisticated machinery and management to achieve higher yields.

Since extensive production technology has predominated in past years, the pond shrimp industry has expanded largely at the expense of coastal resources. Mangrove swamps, which are characterized by extremely high biological productivity and are therefore a critical element of coastal ecosystems, have been displaced. In addition, PL collection has at times been excessive and wastewater emissions from some enterprises have harmed the environment. Mariculture also suffers from water pollution from agricultural, urban, and industrial sources.

In the first half of this chapter, the extent and consequences of coastal ecosystem disturbance are described. After that, a causal analysis of environmental problems is presented. The policies contributing to depletive management of wetlands and related resources are shown to be similar to the policies stimulating encroachment on tropical forests and other natural habitats. For one thing, the tenurial regime rewards those who convert coastal ecosystems into shrimp ponds, just as frontier property arrangements encourage farmers and ranchers to colonize "idle lands" in other parts of the country. In addition, mariculture's geographic expansion, like agriculture's, has been caused in part by inadequate spending on education, research, and extension.

Alteration of Coastal Ecosystems

Damage to coastal resources is the subject of widespread concern in Ecuador. Although many shrimp ponds have been constructed in *salitrales* (i.e., the salt flats that build up through evolution of the mangrove system), maricultural expansion has been considerable in intertidal areas alongside the shore. Particularly troubling has been the displacement of mangrove swamps. Renewable resources also come under threat because of excessive PL capture and water pollution.

Mangrove Deforestation

Remote sensing studies carried out by Ecuador's Center for the Integrated Survey of Natural Resources through Remote Sensing (CLIRSEN) suggest that 118,000 hectares of shrimp ponds had been established along the Ecuadorian coast by 1987. Of that total, 38,500 hectares had been constructed in salt flats and 28,500 hectares directly displaced mangrove swamps. Urban expansion accounted for an additional 1,400 hectares of mangrove deforestation. In all, 15 percent of the mangrove swamps and three quarters of the salt flats existing in Ecuador in 1969 had disappeared by 1987 (Table 8-1).

In some locations, coastal ecosystem disturbance has been more extreme. A little over 30 percent of the mangrove swamps in El Oro province, south of Guayaquil, were displaced between 1969 and 1991. The situation is worse in Manabí province, which has lost nearly two thirds of its mangrove swamps (Sanchez, 1992, p. 16).

Mangrove deforestation accelerated through the early 1990s. More than 21,000 hectares were uprooted from 1969 to 1984. This translates into an annual rate of 0.7 percent. Between 1984 and 1987, swamp

Table 8-1 Area of Mangrove Swamps, Salt Flats, and Shrimp Ponds (ha), 1969, 1984, and 1987

	Mangroves	Salt flats	Shrimp ponds
1969	203,700	51,500	0
1984	182,100	20,000	89,400
1987	175,100	12,400	117,700

Source: CLIRSEN data.

displacement almost doubled, to 1.3 percent per annum, nearly 7,000 hectares having been lost during that three-year period. The deforestation rate reached 1.9 percent between 1987 and 1991, when another 13,000 hectares were destroyed (Sanchez, 1992, p. 16).

Interestingly, the recent increase in mangrove deforestation has taken place in spite of falling shrimp prices (see below) and capital shortages, which make establishing a maricultural enterprise more expensive. Furthermore, many of the enterprises brought on line since 1987 have chosen to locate outside of wetlands because soils in mangrove swamps tend to be highly acidic, which creates problems for shrimp production.

It should be remembered, though, that mangroves are not torn out just to create additional shrimp ponds. The slums of Guayaquil, Machala, and other coastal cities tend to establish themselves in wetlands because, as is noted later in this chapter, those areas are essentially an open-access resource. Even when wetlands are not destroyed entirely, ecosystem degradation occurs because mangrove swamps are a source of charcoal and building materials.

Overfishing

The maricultural industry is becoming more concerned about periodic shortages of the PL used to stock shrimp ponds.

When the ocean off the Ecuadorian coast is warm, marine life is abundant. There are a few thousand full-time PL fishermen, complemented by perhaps 10,000 working a few days a month (Scott and Gaibor, 1992). As long as prices do not fall too low, a skilled gatherer can earn $25 to $30 a day (Thia-Eng and Kungvankij, 1989).

The capture of wild PL involves considerable losses. Of the 9 to 12 billion collected annually during the early and middle 1980s, only half were of the desired species, *Penaeus vannamei*. Other kinds of shrimp were discarded, usually on dry sand where they died. Other captured organisms met the same fate. Mortality of *P. vannamei* was also high, exceeding 80 percent between the beaches and the ponds (LiPuma and Meltzoff, 1985, p. 20).

In a good or ordinary year, these losses have no great effect on Ecuadorian mariculture. But when coastal waters are cool, populations of juvenile shrimp in the wild fall dramatically, which can cause some pond capacity to sit idle. To guard against this risk, the industry began constructing hatcheries in the middle 1980s. Existing hatcheries, which number 120 or so, can satisfy as much as 65 percent of industry demand (R. Barniol, personal communication, 1992).

A few operations, at which mature shrimp are mated repeatedly to produce PL (R. Barniol, personal communication, 1992), are largely insulated from the impacts of overfishing and habitat destruction. But most hatcheries are not self-contained. Instead, gravid females captured by oceangoing trawlers and artisanal fishermen are the primary source of seed. Of course these latter facilities often have to shut down when gravid females become scarce, which occurs when climatic conditions are adverse.

Water Pollution

The environmental impacts of mariculture development are not limited to uprooting mangroves and collecting too many PL and gravid females in the wild. Wastewater discharged from shrimp ponds can impair coastal water quality, which is already threatened by pollution from urban, industrial, and agricultural sources.

Many maricultural operations around the Gulf of Guayaquil are semiextensive, with pond stocking rates ranging from 10,000 to 50,000 PL per hectare (Villalón et al., 1989, p. 251). At those rates, supplementary fertilization, with urea and superphosphates, is needed to induce the phytoplankton blooms on which juveniles feed. Fertilization, along with supplemental feeding for larger stock, depletes oxygen. If dissolved oxygen levels fall too low, shrimp die.

To avoid this outcome, seawater is exchanged, through pumping, for pond water. Water flushed from ponds contains nutrients, which increases biological oxygen demand in surrounding waters and can accelerate eutrophication. Twilley (1989, p. 98) suggests that this might contribute to the red tides (blooms of red-colored toxic algae) observed occasionally in the Gulf of Guayaquil. He also speculates that daily discharges of semiextensive operations' wastewater, which can be highly saline (Snedaker et al., 1986), into the Gulf of Guayaquil might exceed freshwater yield from the Guayas River Basin during the dry season (Twilley, 1989, p. 98).

Aside from what takes place in habitats adjacent to maricultural enterprises, the environmental impacts associated with discharging wastewater from ponds probably pale in comparison with water pollution from other sources. Solórzano (1989) finds that emissions of untreated domestic and industrial waste from the City of Guayaquil are a principal cause of high bacterial contamination, enhanced nutrient concentrations, and low dissolved oxygen content in the Guayas River, which is the principal freshwater tributary of the Gulf of Guayaquil. Water pollution around other coastal cities is, likewise, severe. Solórzano (1989) also reports high nitrate and pesticide pollution from agricultural sources. Finally, freshwater flow into the gulf during the peak of the wet season has been reduced significantly since completion of the Daule-Peripa Dam (Arriaga, 1989, p. 151). However, the same project has probably caused freshwater flow to increase during the dry season, when salinity and pollution are probably more serious problems.

Consequences of Coastal Ecosystem Disturbance

Determining the consequences of mangrove deforestation, water pollution, and overfishing is difficult. The life cycle of *P. vannamei* in the wild has been characterized. Specifically, it is known that females shed their eggs in the open sea, where they hatch. After a larval phase, PL move into estuaries seeking out nutrient-rich niches (e.g., those around mangrove roots). As they mature, they become bottom-dwelling and, after a few months, they return to the sea (McPadden, 1985). In addition, the environmental factors (e.g., temperature and salinity) influencing this cycle are generally understood.

However, much more biological research must be done before the economic costs associated with coastal ecosystem disturbance can be estimated. Among the empirical questions that need to be answered are those that follow.

> How are natural stocks affected both by PL harvesting and by collection of gravid females?
> What is the relationship between PL numbers and the extent and quality of mangroves and other coastal habitats?
> What impacts does water pollution have on plant and animal populations in the wild?

Research conducted in other parts of the world suggests that linkages addressed by the preceding questions are not trivial. In particular, positive relationships between coastal wetland area and shrimp catch have been documented in Malaysia, the Philippines, and the Gulf of Mexico (Turner, 1989, pp. 123–124).

In addition, economic evidence suggests that habitat destruction and overfishing have had adverse impacts on shrimp mariculture in Ecuador. By the middle of 1988, prices paid by hatcheries for gravid females had risen to $10, according to industry sources. By late 1991, prices were ranging from $30 to $55 when fishing for gravid females was legal and up to $120 when temporary bans on such fishing were in effect. The price increase probably reflects a decline in natural supplies.

In real terms, the price paid by maricultural enterprises for PL has risen from $2 a thousand in the late 1970s to over $4 a thousand throughout the 1980s. Those prices vary considerably, however. During years like 1983, when there is a strong El Niño climatic event, PL are abundant because shrimp stocks rise rapidly along with ocean temperatures. At other times, prices can exceed $10 a thousand (Sutinen et al., 1989, p. 40). In 1985, when prices peaked at $15 a thousand, payments for PL amounted to 44 percent of the cost of operating a semiextensive enterprise and PL shortages resulted in underutilization of pond capacity (LiPuma and Meltzoff, 1985, pp. 17–18). Many ponds were idled in 1990 for the same reason.

In 1993, a new threat to shrimp production emerged. Responding to the spread of Black Sigatoka fungus, the owners and managers of banana plantations have increased spraying of pesticides dramatically in recent years. Substances that are banned or tightly controlled in the United States are being used with few precautions, with applications many times greater than what manufacturers recommend. As a result, toxic substances are working their way into shrimp ponds downstream.

Maricultural enterprises in the lower Guayas River Basin are convinced that their production is falling because of pesticide pollution. High mortality among juvenile shrimp and stunted growth among survivors, which seems to result from exposure to pesticides, have been reported and total national output declined by more than 10 percent between 1992 and 1993. At the request of the Ecuadorian government and the IDB, the U.S. Environmental Protection Agency (EPA) will send experts to the country in 1994 to assess the threat to the shrimp industry associated with agricultural chemical run-off.

Causes of Environmental Degradation along the Ecuadorian Coast

Like other parts of Ecuador's rural economy, the shrimp industry suffers because the government has no effective policies for dealing with environmental externalities, like pesticide pollution (see above). In addition, it feels the impacts of governmental interference with market forces. Exports are subject to a 1 percent tax. Also, imports of high-quality feed have been restricted at times, to the detriment of maricultural productivity (Rosenberry, 1990).

Overvaluation of the sucre can also discourage production and impinge on industry earnings. However, the main effect is to encourage smuggling. In 1984, when the official exchange rate was only 80 percent of the market rate, a fifth of the Ecuadorian shrimp harvest was shipped illicitly to Peru, from where it was sent on to the United States and other countries (LiPuma and Meltzoff, 1985, p. 20).

Since smuggling is relatively easy, the impacts of currency distortions on shrimp industry performance are not all that great. Two other elements of the policy environment have a much stronger influence on maricultural enterprises' use and management of coastal resources. The first is property arrangements. The second element is inadequate investment in the shrimp industry's scientific base.

Inappropriate Property Arrangements

Depletive management of Ecuador's coastal ecosystems has much to do with the legal standing of resources. By law, coastal beaches, saltwater marshes, and everything else below the high-tide line is a national patrimony. But access to most of that land is completely free. For example, no public agency attempts to keep an accurate count of PL collectors in different parts of the country. Never, it seems, has there been a serious proposal to subject that group's activities to legal control.

Permanent occupation of coastal wetlands is also unregulated. For example, expansion of the slums of Guayaquil and Machala into adjacent wetlands is totally uncontrolled. Indeed, some politicians have encouraged that land use change.

There is some regulation of shrimp pond construction along the shore. Specifically, a ten-year use permit must be obtained from the General Merchant Marine Directorate (DGMM). Depending on the pond's location, approvals might be needed from INEFAN, IERAC, and other public agencies. Annual fees charged permit holders amount to 11 percent of the official minimum wage for each hectare. Generally, this works out to less than $10 per hectare per annum.

Some individuals have been able to construct ponds without permits and, once operations have begun, to claim that the site was above the high-tide line (and therefore not subject to public control). Mariculture is even being pursued, without any sort of governmental approval or interference, in the Churute Ecological Reserve, 40 kilometers south of Guayaquil. Other entrepreneurs have found it useful to take on a government official as a partner. The advantage of doing so is that the many months normally spent waiting for a permit to be approved can be avoided. Another option is to offer bribes, which are reported to have reached $100 a hectare in the middle 1980s (LiPuma and Meltzoff, 1985, p. 9).

The tenurial roots of coastal ecosystem degradation can be described as a mixture of a "tragedy of the commons" and rent capture. Excessive PL collection along beaches is, like overfishing in the ocean, a clear example of overexploitation of an open-access resource. The benefits of extra fishing effort (i.e., the value of additional catch) are private. By contrast, the costs associated with decreased breeding populations and other forms of fishery depletion are shared by all who make their living, directly or indirectly, from coastal ecosystems.

Water pollution from shrimp ponds and other sources is, similarly, a tragedy of the commons. The benefits of releasing saline water rich in nutrients into public waterways are internalized by the individual operator (in the form of avoided treatment costs) while the costs of emissions (associated with damage to ecosystems) are an externality.

Habitat destruction involves negative externalities as well. However, the conversion of wetlands and other coastal ecosystems is also an example of rent capture. A rudimentary analysis shows that the net revenues generated by a shrimp pond far exceed the costs associated with obtaining use permits. For a semi-intensive operation yielding only 1.80 metric tons per hectare per annum, average per-hectare costs are around $6,000 a year (LiPuma and Meltzoff, 1985, p. 17). With shrimp producers currently receiving more than $4,000 a metric ton for their output, annual net returns can easily reach $2,000 a hectare. Clearly, a significant portion of coastal ecosystem destruction has been motivated by individuals' desire to capture that income stream in exchange for annual fees of less than $10 a hectare.

Inadequate Investment in Human Capital and the Scientific Base

The property arrangements contributing to the conversion of coastal ecosystems into shrimp ponds are similar to those that accelerate agricultural colonization of other natural habitats. There is another parallel between mariculture's geographic expansion and agriculture's, which is that both have been a consequence of inadequate investment in human capital, research, and extension.

The history of the Ecuadorian shrimp industry is a classic illustration of Hayami and Ruttan's (1985) thesis that geographic expansion usually precedes productivity-enhancing investment in agriculture and other parts of the rural economy. The first ponds, in which extensive technology was employed, were built close to shore. Dikes by the sea could be opened to let in clean seawater and the PL and nutrients it contains. Yields were minimal. But costs were also low, there being no requirement for supplementary stocking, fertilization, or feeding.

As the sites best suited to extensive mariculture have been occupied, production technology has changed. Ponds located farther from and a little above the ocean have to be stocked and fertilized artificially. As is noted earlier in the chapter, this makes mechanized exchange of pond water for water from the sea or estuaries necessary. The transition to semiextensive technology has also been accelerated by subsidization of the diesel fuel used to operate pumps. Had Ecuadorian energy prices been at international levels during the middle 1980s (BCE, 1990), spending on diesel fuel would not have accounted for 12 percent of the operating costs of a semiextensive maricultural enterprise. Instead, fuel expenditures would have amounted to more than a quarter of those costs (LiPuma and Meltzoff, 1985, p. 17).

Although spending on pumps and other machinery can be considerable, management of many semiextensive operations is still rudimentary. Stocking, fertilization, and application of antibiotics (to combat disease) are usually haphazard. In addition, many enterprises neither employ biologists nor maintain records that would, over time, allow them to produce more efficiently.

In recent years, semi-intensive technology has begun to be adopted. Ponds are being designed and operated to achieve efficient water exchange and aeration. The number of biologists employed to manage PL stocking and application of feed and nutrients is also increasing. In addition, stocking densities in semi-intensive ponds are higher, as are yields.

Further improvement in shrimp production technology can be expected in Ecuador, with more enterprises undertaking the careful management that characterizes a semi-intensive operation. However, aside from firms located close to Guayaquil, the industry's access to laboratories where water quality can be tested and where shrimp diseases can be assessed remains minimal. Accordingly, opportunities to achieve more precision in the application of fertilizers, feed, and antibiotics and also in water exchange are not being fully exploited.

Additional spending on research and technology transfer would also ease Ecuadorian mariculture's dependence on PL captured in the wild. In particular, more widespread understanding of the factors influencing shrimp reproduction (i.e., temperature, water chemistry, and above all nutrition) would allow more hatcheries to become self-contained. Once this is done, the problem of excessive collection of PL and gravid females will be solved.

The Future of Ecuadorian Mariculture and the Country's Coastal Ecosystems

To be sure, many maricultural enterprises in Ecuador have adopted production technology that is capital intensive, involves more sophisticated biological management, or both. Nevertheless, extensive operations, featuring low costs and low yields, continue to be the norm. As of 1987, some 60 percent of Ecuador's shrimp ponds were extensive, 25 percent were semiextensive, and only 15 percent were semi-intensive (CPC, 1989, p. 27).

Because extensive production technology is still dominant, shrimp yields in Ecuador, which averaged 0.59 metric ton a hectare in 1987 (Table 8-2), are below world norms. They are only 71 percent of what is achieved in Honduras, which is the second largest producer in the Western Hemisphere. In Mexico, which recently reformed the investment

Table 8-2 Maricultural Productivity, World's Top Five Shrimp Exporters

	Production (t)	Yield (t/ha)
China[a]	165,000	1.06
Indonesia[a]	90,000	0.36
Thailand[a]	90,000	1.12
Ecuador[b]	70,000	0.59
Philippines[a]	50,000	0.62

[a] Production and yield in 1989 from Rosenberry (1990).
[b] Production and yield in 1987 from General Fishing Directorate (DGP) reports and Table 8-1.

and land acquisition policies that prevented it from growing shrimp for the lucrative North American market (where most of Ecuador's output is shipped), annual yields are 1.00 metric ton a hectare (Rosenberry, 1990). Even more ominous in terms of Ecuador's international competitiveness is the performance of major Asian producers. As is reported in Table 8-2, Chinese and Thai yields are two thirds higher.

The entry of more efficient producers into the world shrimp market has caused prices to fall. The average value Ecuador received for its exports declined from $10,671 to $6,439 (in 1990 dollars) per metric ton between 1986 and 1990. The decline has been even more pronounced for maricultural enterprises. Whereas market values for larger shrimp captured off the coast have stayed relatively stable, values for the smaller product yielded by ponds have fallen dramatically as additional countries have entered the market.

One possible consequence of growing international competition could be to put less efficient producers out of business. Some extensive operations ought to survive because of their low costs. Also, semi-intensive enterprises, which have relatively high yields, can probably withstand further price declines. The threat to semiextensive operations, though, is more serious. Their costs are higher than those of extensive producers while their yields are not all that great. Additional reductions in the value of output will be difficult for them to bear.

Tough times for some parts of the Ecuadorian shrimp industry could result in environmental damage. Producers with poor long-term prospects will not hestitate to damage coastal ecosystems if an economic rent can be captured by doing so. In particular, they are not likely to adopt pollution controls. Similarly, they will resist any initiative to reduce overfishing if short-term PL prices are driven up as a result.

Finally, maricultural development as a whole will continue leading to ecosystem destruction if property arrangements are not reformed. As long as some segments of the industry remain profitable, it will be possible for individuals to capture rents by converting mangrove swamps and other public lands into shrimp ponds. Unless the government stops treating those lands as a free good, they will be destroyed.

9

Tourism and Species Preservation in the Galápagos

To many foreigners, Ecuador is best known for the Galápagos Islands (Figure 9-1), which are located about 1,000 kilometers west of the South American coast. Being relatively isolated, the archipelago is home to a unique mix of flora and fauna. Of course, anybody who has taken a biology class knows of Charles Darwin's brief visit to the islands, described in a highly entertaining chapter of the *Journal of the Voyage of the Beagle* (1845). The observations he made there eventually flowered into his theory that new species arise because of natural selection (Darwin, 1859).

By the time of Darwin's voyage, the Galápagos were on their way to being a lost Eden. Indigenous species have few defenses against predatory animals and competing vegetation introduced since the islands were discovered. Some native animals, including the giant tortoise for which the archipelago is named, were nearly hunted to extinction.

Since time might be running out on local wildlife, visitors are flocking to the islands. This has its good and bad points. Tourism can generate the funds needed for conservation programs. But in the absence of suitable controls, people can damage the ecosystems they wish to experience firsthand.

At the beginning of this chapter, past human intervention in the archipelago and recent conservation initiatives are reviewed. Next, growth in tourism in the Galápagos is documented. We argue that, because visits have been undertaxed, the potential for environmentally

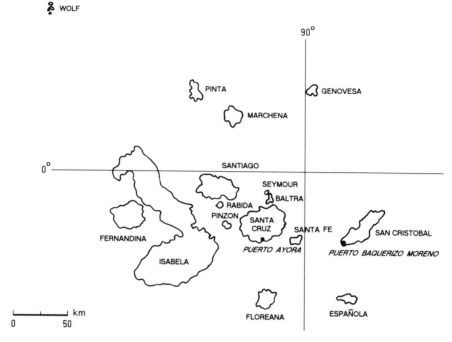

Figure 9-1 The Galápagos.

destructive tourism has been enhanced and an important source of support for species and habitat management, which is desperately needed, has not been exploited.

A Short History of the Galápagos

There are about 800,000 hectares of land in the Galápagos, most of which is accounted for by five islands—Isabela, Santa Cruz, San Cristóbal, Santiago, and Floreana (Figure 9-1). The archipelago is of recent geologic origin, the oldest lavas dating back only 3 to 5 million years. Eruptions and other volcanic events continue to occur frequently.

The islands are populated by a truly unusual assemblage of flora and fauna. Among the noteworthy species are gigantic land tortoises, prehistoric-looking marine and land iguanas, numerous marine and land birds, as well as sea lions and fur seals, all of which show little or no fear of humans. About 60 percent of all plants and animals are endemic (de Groot, 1983). Remarkable groups of Galápagos organisms, such as thirteen forms of Darwin's finches, provide textbook examples of rapid evolution.

Based on an expedition he led in 1953, Norwegian explorer Thor Heyerdahl argued that voyages from South America to the archipelago took place in pre-Incan times. The first human contact since the beginning of recorded history in the New World was in 1535, when the ship carrying Bishop Tomás de Berlanga and his party from Panama was becalmed and carried far from the American mainland. For many years after this sojourn, buccaneers and other adventurers were the only people to make sporadic visits. The Galápagos have belonged to Ecuador since 1832, when an expedition was sent out by Ecuador's first president, General Juan José Flores.

By the time Darwin stopped over for a few weeks in 1835, it was common practice for whaling vessels and other merchant craft to pass by the islands to pick up tortoises, which could stay alive for up to a year in a ship's hold. Since there were few other ways to have fresh meat during a long voyage, tortoise hunting continued for several decades, which caused some island populations to disappear.

Because of recent volcanic activity and a general lack of fresh water, no more than a tenth of the soils in the archipelago are suitable for agriculture, according to resource assessments conducted by PRO-NAREG. Nevertheless, Ecuador has made several attempts to settle farmers and ranchers. In the late 1800s, convict laborers were brought in to establish sugar plantations on Floreana and San Cristóbal, which are the only two places where freshwater supplies are fairly dependable. These initiatives came to an end when workers rebelled and killed their overseers.

In the early twentieth century, the Galápagos had practically no human settlers and just a few visitors. Attempts to collect *orchilla* (a dye-yielding lichen), salt, and guano fared poorly. A Norwegian fishing and canning enterprise, launched in 1925, was abandoned in 1928.

The years immediately before World War II saw the arrival of a motley array of Europeans. The islands acquired notoriety for eerie romance after the unexplained disappearance, in 1934, of an ersatz baroness and one of her lovers, who had settled on Floreana. Attracted by the Galápagos's mystique, wealthy American yachtsmen made intermittent calls, creating a market of sorts for the vegetables and livestock produced by the few year-round residents.

From 1942 to 1948, the U.S. military operated an air base on Baltra to protect against submarine attacks on the Panama Canal. Right after World War II, Ecuador established a penal colony on Isabela. The inmates' survival often depended more on tortoise and bird hunting than on government-supplied provisions. The colony was officially disbanded in 1959.

Conservation Initiatives

Attempts to settle the Galápagos, which at one time or another involved farmers, prisoners, romantics, and soldiers, generally failed to last more than a few years. By contrast, the environmental impacts of human incursions have proven to be permanent.

Mainland organisms probably began to establish themselves in the archipelago when the first ships made a landfall. The widely circulated notion that goats and pigs were introduced by sailors hoping to guarantee a food supply during subsequent visits may by apocryphal. However, it is undeniable that domestic animals gone feral and rats have adapted readily to the local environment. In addition, indigenous plants have often lost ground to flora brought in, purposefully or inadvertently, from the mainland (Jackson, 1990).

Introduction of exotic organisms, which continues to this day, has thrown island ecosystems far out of equilibrium. If the Galápagos were simply left alone, many more indigenous species would become extinct. To avoid the outcome of a desolate landscape populated only by goats, rats, and cats, corrective action is required.

Ecuador first showed interest in protecting Galápagos wildlife during the centennial of Darwin's visit. The first two conservation laws ever passed in the country related directly to the archipelago. A nature sanctuary was established there in 1934 and hunting of selected island species was prohibited two years later. However, the islands were too remote for these laws to be enforced.

International interest in the fate of the archipelago was catalyzed by the United Nations Educational, Scientific, and Cultural Organization (UNESCO), one of the founders of which, Sir Julian Huxley, was a great-grandson of Darwin's foremost supporter. After alarming reports issued in the middle 1950s, contacts were made between UNESCO and Ecuadorian officials in Europe. This resulted in a UNESCO-sponsored team being sent to the Galápagos in 1957 to prepare for establishment of a research station that would take charge of protecting endangered species.

This initiative gained support at the Fifteenth International Zoological Congress, held in London in 1958. At that congress, a prestigious group of scientists, including Huxley, issued a statement declaring that conserving the Galápagos was urgent and required international collaboration. The International Union for the Conservation of Nature (IUCN) responded to this plea by taking the lead in organizing a rescue campaign.

The first tangible result of the campaign was the creation, in Brussels in July 1959, of the Charles Darwin Research Foundation. The same year, President Camilo Ponce of Ecuador gave the foundation official authorization to operate in the Galápagos for twenty-five years. (That authorization has since been renewed.) Its Charles Darwin Station, located on the outskirts of Puerto Ayora (Figure 9-1), was dedicated in January 1964. The station coordinates scientific research in the islands. It has also had considerable success with its programs to revive threatened land tortoise populations.

Establishment of the Darwin Station, combined with growing national interest in nature reserves, led to the creation of Galápagos National Park, the limits of which were demarcated in 1969 and 1970. Except for a little more than 3 percent of the land area already occupied by the military, towns, and farms, the entire archipelago lies inside the park. The first superintendent was appointed in 1972.

The Growth of Tourism

Tourism has followed close on the heels of modern scientific expeditions. The Galápagos International Scientific Project of January and February 1964 demonstrated the feasibility of bringing relatively large groups to the islands. Two boats, the USS *Pine Island* and the California Maritime Academy *Golden Bear,* transported sixty-six scientists from San Diego, California. Simultaneously, the Ecuadorian Air Force flew government officials, diplomats, students, and special guests from the Ecuadorian mainland for the dedication of the Darwin Station. After this event, a small but steady stream of visitors was attracted by the station's presence.

In the late 1960s, two Ecuadorian travel companies, Metropolitan Touring and Turismundial, were contacted by Lars Eric Lindblad and other established cruise operators interested in bringing groups to the Galápagos. At the same time, the availability of larger aircraft to the Ecuadorian Air Force opened up the possibility of flying passengers to the vacant airfield on Baltra. The boat operators soon linked up with the military running the flight facilities and the era of what would later be called ecotourism began.

According to data collected by the Galápagos National Park Service (SPNG), numbers of visitors have increased steadily during the past twenty years, from fewer than 5,000 in 1970 to more than 40,000 currently. Demand for trips to the Galápagos has been fed by an increase in nature-related film and television documentaries, which has in turn made many individuals keen to see the islands for themselves (Machlis et al., 1990). Furthermore, there is hardly a school in the western world where Darwin's evolutionary ideas are not discussed, which creates even more public interest.

Tourists tend to be affluent and well educated. In a survey of 457 visitors carried out in July and August 1990, Machlis et al. (1990) found that 70 percent were from North America or Europe and that three quarters had at least some university education. Average income was nearly $37,000 (Edwards, 1991). Less than 15 percent of the Machlis et al. (1990) sample spent more than seven nights in the archipelago. The vast majority of visitors travel and stay on boats, either the cruise vessels holding up to ninety passengers or smaller craft, most of which provide their six to twelve passengers with more flexible schedules.

SPNG data indicate that the number of Ecuadorian tourists has risen dramatically since the middle 1980s. This could have an impact on economic activity on land because, unlike foreign tourists, Ecuadorians prefer to stay ashore in hotels. In addition, there is a recent trend toward shore-based tourism for foreigners, who can now stay in nicely appointed hotels and take day trips on fast boats to different sites.

Expanded tourism has stimulated infrastructure development. The airport on Baltra has been improved and a road now runs from it across Santa Cruz to Puerto Ayora, which is a tourism hub. Another airport was opened in 1986 next to Puerto Baquerizo Moreno (Figure 9-1), the provincial capital on San Cristóbal island.

A major consequence of tourism and associated infrastructure development has been rapid immigration from the mainland. When the Charles Darwin Station was dedicated in 1964, fewer than 3,000 people resided in the Galápagos. Once the archipelago was made a province in 1973, no restrictions could be placed on Ecuadorians wanting to travel in or out or to stay. Perceiving that jobs and business opportunities are plentiful in the islands, thousands have taken up permanent residence, mainly in Puerto Ayora and Puerto Baquerizo. Of the 9,785 individuals counted in the 1990 census, 7,317 lived in the two towns (INEC, 1991, p. 32).

The factors pulling migrants into the Galápagos are strong. A relatively high portion of the islands' households is connected to potable water and sewer systems and also have electricity (Table 9-1), although the quality of public utilities is not always satisfactory. Prices for food and other consumer goods, nearly all of which are imported from the mainland, are high. But salaries are also above continental norms and unemployment is relatively low.

Tourism Pricing Issues

Tourism can damage fragile island environments. If visitors are allowed to wander at will, soils can be compacted, erosion can accelerate, or both. Nesting sites for birds can also be disturbed. In addition, jettisoning of solid wastes from boats is a potential threat to marine and coastal wildlife and tourists' purchases of souvenirs made from black coral could cause it to disappear from the Galápagos.

Mountfort (1974) reports that careless photographers occasionally interfere with the breeding of birds and de Groot (1983) complains of people chasing marine iguanas, which must lie quietly on warm rocks to

Table 9-1 Selected Quality-of-Life Indicators in 1990 for the Galápagos and Ecuador

	Percentage of households with		
	Potable water	Sewage service	Electricity
Galápagos	88.6	90.3	94.8
All of Ecuador	62.7	72.3	77.6

Source: INEC (1991), p. 223.

recover body heat after emerging from the cold ocean. But in general, tourists exercise caution as they go about their business. Relatively few hikers wander off clearly marked paths. In addition, most visitors do not buy anything made of black coral. All told, the adverse impacts of tourism are probably minuscule compared to the ecological imbalances that are the lasting legacy of past human incursions.

Ostensibly to help maintain environmental quality, the Ecuadorian government indirectly limits the number of visitors to the Galápagos. Passenger boats cruising the islands must be licensed. In addition, official permission has to be secured before a hotel can be constructed.

The regulatory approach has serious drawbacks. For one thing, individual companies have strong incentives to overcome bureaucratic hurdles in order to enter the Galápagos tourism market, which can be highly lucrative. Furthermore, by regulating tourism instead of taxing it, the Ecuadorian government has failed to collect funds that could be used to maintain (or, better still, to revive) natural ecosystems.

The earnings of Galápagos tourism operators can be determined by inspecting SPNG's comprehensive data on where boats licensed to cruise the archipelago actually took their passengers and also information on prices paid by foreign and national tourists. Using these data, Bruce Epler (personal communication, 1992) found that gross earnings for large cruise ships were around $4 million each in 1990. Of course, expenditures on fuel, crew wages, food, and so forth, would have to be subtracted to yield income. Unfortunately, data on those expenditures are not available.

Although net, as opposed to gross, earnings of tourism operators cannot be estimated, one can be reasonably sure that annual fees paid to the government for the right to cruise the Galápagos have been a negligible portion of income. Payments by any individual enterprise depend on the passenger capacity of its craft. In 1991, for example, SUFOREN collected around 9.74 million sucres, which was roughly equivalent to $9,000, from sixty-one boats and ships, which had a combined capacity of 844 passengers. This works out to $10 a berth. Even the largest boats in the islands, which can carry as many as 90 individuals, not counting the crew, paid only 900,000 sucres. This amount is probably less than or equal to the profits earned in a month or two by such a boat's gift shop and bar. (Fortunately, INEFAN, which replaced SUFOREN as the principal authority for park management in Ecuador, raised fees to $200 a berth in early 1993.)

The impacts of underpricing access to the Galápagos are entirely predictable. For one thing, private firms' earnings have been enhanced. For example, the few airlines licensed to service the islands are able to charge foreigners high prices; round-trip fares from the mainland currently range from $350 to $400. Furthermore, one firm or another, motivated by the prospect of capturing more rents, is always lobbying to bring a ship out to the islands or to build a new shore facility. This behavior is no different from logging or settling in underpriced tropical forests (Chapter 4) or locating a shrimp pond in coastal wetlands in response to the government's policy of charging only a pittance for use permits (Chapter 8).

Research undertaken by Edwards (1991) provides a glimpse into the financial contribution that visitors could make to conservation initiatives in the archipelago. The purpose of his study was to estimate the maximum amount that visitors would pay for Galápagos trips. Virtually all foreigners stop at one or more places on the South American mainland (e.g., Quito, the Amazon, or Machu Picchu) on their way to or from the islands. Indeed, most tour packages oblige them to do so. Under these circumstances, standard travel cost methodology cannot be used. That is, a trip's value cannot be inferred simply from what visitors pay in the islands and for air tickets. As an alternative, Edwards (1991) undertook a hedonic demand analysis. With this technique, the number of visits to different places as well as the time spent at various destinations are taken to be endogenous and the marginal value, or imputed price, of time spent at any particular site is estimated on the basis of a comprehensive review of expenditure patterns (Palmquist, 1991).

Applying this approach to data collected in a 1986 survey of 250 tourists, Edwards (1991) estimated that the implicit price of time spent in the Galápagos is a little more than $300 a day. He also investigated the consequences of restructuring park entrance fees, which were recently raised to $80 a visit. He found that, if a daily fee of $214 were paid, the average length of visits would fall by half, because the marginal value of time spent in any one place, including the Galápagos, is a diminishing function of the length of the visit. However, nearly twice as many tourists would arrive, thereby allowing the total number of visitor days, which is subject to indirect controls, to stay constant. In addition, tax receipts would rise by a factor of twenty or more.

The study has limitations. For one thing, it seems to rest on the assumption that there is a strict limit on the number of tourists allowed into the Galápagos. Also, differences between expenditures by Ecuadorians and foreigners do not appear to have been addressed clearly. For example, the park entrance fee paid by Ecuadorian visitors was only 600 sucres (equivalent to less than a dollar in 1992) up to the beginning of 1993 and their round-trip airfares continue to be around $150. In addition, the impact of reducing the average length of visits on tourism operators' costs was not investigated. If those costs were driven up by a reduction in average trip length, operators could be expected to resist the proposal to link visitor's fees to time spent in the islands. In spite of these shortcomings, however, Edwards's (1991) research clearly shows that tourism in the Galápagos has been significantly underpriced.

Conclusions

In general, tourism does not directly threaten the Galápagos Islands' fragile ecosystems. However, the possibility that the archipelago's primary industry could grow to environmentally unsustainable dimensions arises because tourism operators and their clients are not paying their full way. In particular, fees collected from operators are extremely small, which gives them a strong incentive to overcome indirect regulatory controls on their operations. Recent years' growth in the number of visitors suggests that those controls are not very effective.

Raising fees could provide the financial resources needed to address the most serious environmental problem in the Galápagos. Far from being pristine and harmonious, the islands' ecosystems have been thrown far out of equilibrium by the flora and fauna introduced, intentionally or by accident, by humans since the sixteenth century. To nurture those ecosystems (so as to avoid continued species extinction) will require the funds that could be supplied by an adequately taxed tourism industry.

III

CONCLUSIONS AND RECOMMENDATIONS

10

Development and the Environment: Some Common Fallacies

This book's analysis contrasts sharply with some widely circulated notions of why environmental quality is declining in places like Ecuador. Many believe that demographic growth is out of control and, as a result, the human population has already or will soon exceed what the environment can support. Others say that economic growth, pure and simple, needs to be kept in check because it puts too much pressure on natural resources.

Still others view the unfettering of market forces in Latin America with concern. They concede that this departure from past economic policy is needed to stimulate development. However, they worry that the environment and poor people will suffer. Yet another perspective has nothing to do with Malthusian pessimism. Neither do its adherents necessarily question economic liberalization. Instead, they doubt that the investment in education and technological improvement required to achieve sustainable development is within poor countries' reach. In particular, they point to the debt crisis and adverse terms of trade as reasons why fiscal capacity is unduly restricted in Africa, Asia, and Latin America.

If valid, any one of these arguments should cause a thinking person to worry about the future of most of the developing world. Unfortunately, none can be comprehensively refuted for the simple reason that each contains elements of truth. Demographic and economic growth can exceed what natural conditions can sustain. Externality problems arise and poverty exists in economies where market forces are given free play. Debt problems and declining prices for exports have caused Africans, Asians, and Latin Americans to suffer. As a result, many of them have had no choice but to pick renewable resources clean.

But environmental degradation and poverty need not be Ecuador's fate. The evidence and the counterexamples offered in this chapter demonstrate that material standards of living can rise dramatically through the wise use and management of renewable resources. The policy reforms needed to achieve environmentally sound economic development are identified in the next, and final, chapter of this book.

Is Population Growth Out of Control in Ecuador?

Had this book been written ten years ago, the argument could well have been made that a Malthusian scenario was being played out in Ecuador. Dramatic increases in life expectancy observed since the 1950s had not been matched by significant reductions in human fertility. As a result, the population was increasing by nearly 3.0 percent a year—in what was already South America's most crowded country (Chapter 2).

To contend in 1983 that rapid population growth was going to continue indefinitely would have been to doubt Ecuadorians' ability to respond to changed circumstances by having fewer children. As is documented in Chapter 2, urbanization, income growth, improved education for women, and increased availability of family planning services have indeed had a major impact on birth rates, in Ecuador and throughout Latin America.

In light of recent experience, projecting future demographic trends is fraught with pitfalls. Even the World Bank's projections for the next few years are probably inaccurate because they do not seem to reflect recent changes in human fertility.* For example, Bulatao et al. (1990, p. 134) claim that Ecuador's population is currently growing by more than 2.3 percent a year while analysis of returns to the 1990 census suggest that the current rate of annual increase is no more than 2.0 percent (Chapter 2). The projection that 19.60 million people will inhabit the country in 2030 (Bulatao et al., 1990, p. 134) is almost certainly too high.

Of course, it will take some time before Ecuador's population stabilizes. The number of women of childbearing age will increase for several more years (Chapter 2), which means that the country's population will continue to grow well past the turn of the century. However, recent declines in birth rates have given the country more time to adapt to higher population densities. By the same token, the chances have been

---

*In millions of inhabitants, 13.18 in 2000, 15.18 in 2010, 17.67 in 2020, and 19.60 in 2030 (Bulatao et al., 1990, p. 134).

reduced that desperation will eventually drive Ecuadorians to run through all the natural resources at their disposal.

Will Population Growth Outstrip What the Environment Can Support?

Notwithstanding recent declines in the rate of demographic growth, it is all but certain that 20.0 million people will be living in Ecuador four to six decades from now. A question to be faced, then, is whether the country has enough natural resources to feed, clothe, and house a population of this size. Assuming that mismanagement does not occur on a colossal scale, we have no doubt that it has.

As is pointed out in Chapter 6, there are a few semiarid places in Ecuador and precipitation and stream flow vary seasonally in many parts of the Costa and Sierra. But there is no underlying water shortage in the country. To the contrary, its hydrologic resources are the envy of many other nations.

In spite of recent agricultural frontier expansion, Ecuador's standing forests are still extensive and hold large volumes of commercial timber. Furthermore, establishing tree plantations on deforested land, much of which is either partially or fully abandoned, appears to be commercially viable (Southgate, Chase, and Hanrahan, 1993). Ecuador could easily sustain much higher levels of timber production.

Knowledgeable visitors from around the world see myriad possibilities for additional maricultural and fishery development in the country (Thia-Eng and Kungvankij, 1989).

True, farmers and ranchers are settling in areas that are poorly suited to agriculture (Chapter 4). But this is not to deny that many soils, especially those deposited over time in the Guayas River Basin and in highland valleys, are highly fertile.

The notion that good land is unusually scarce is quickly dispelled by a simple international comparison. To facilitate that comparison, two ratios have been calculated for Ecuador and a few other countries. One is the area planted to annual and perennial crops in 1987 (FAO, 1989, pp. 49–54) divided by 1990 population. The other ratio is the area covered by soils that are free of serious physical or chemical limitations (WRI, 1990, pp. 286–287) divided by the same denominator.

Presented in Table 10-1 are the two land resource indicators for Costa Rica, which is generally reckoned to have Central America's strongest agricultural economy, and Colombia, Ecuador's northern neighbor. Also included in the sample are three Southeast Asian nations, two of which have experienced rapid agricultural development in recent years. Certainly by sub-Saharan Africa's standards, crop and livestock production in Zimbabwe is a success, in spite of low output during the drought of the early 1990s.

Table 10-1 Ecuadorian Land Resources: An International Comparison (ha/person)

	1987 Cropland[a] divided by population[b]	Prime Farmland[c] divided by population[b]
1990 Population		
Colombia	0.17	0.17
Costa Rica	0.19	0.03
Ecuador	0.25	0.19
Indonesia	0.12	0.05
Philippines	0.13	0.01
Thailand	0.36	0.02
Zimbabwe	0.28	0.09
2030 Population		
Ecuador	0.14	0.11

[a] FAO (1989), pp. 49–54.
[b] Bulatao et al. (1990).
[c] WRI (1990), pp. 286–287.

The current ratio of cropland to population is nearly twice as high in Ecuador as it is in either Indonesia or the Philippines and exceeds the ratios for Colombia and Costa Rica by a comfortable margin. Of the countries listed in Table 10-1, only Thailand and Zimbabwe have more cropland per capita. In addition, the high quality of Ecuador's land endowment is made clear by the second series of indicators. Relative to the size of its population, no country in the select sample has more soils that are free of major limitations for agricultural production.

Even if projections of population growth made by Bulatao et al. (1990) turn out to be right on the mark, Ecuador's land scarcity problems will be no worse in 2030 (Table 10-1) than what many nations face right now. If the numerator of the first ratio were not to change, per capita cropland would still be a little higher than the current ratios for Indonesia and the Philippines. Other than Colombia, no country in the sample can currently match what per capita endowments of prime farmland will be in Ecuador forty years from now.

Renewable resource degradation occurs in each of the countries listed in Table 10-1. However, few people would say that tropical deforestation, soil erosion, and water resource depletion in Colombia, Costa Rica, Indonesia, the Philippines, Thailand, and Zimbabwe are problems without solutions. Furthermore, in many instances, it has been possible to raise agricultural production enough to feed expanding populations and also to increase exports. These accomplishments cannot be explained only in terms of accelerated depletion of renewable resources.

It is nonsense, then, to think that Ecuador must inevitably reach dire Malthusian straits, in which the population is incapable of feeding itself and exporting agricultural commodities simply because there are not enough renewable resources. That outcome is likely, however, if the country continues to make poor use of those resources due to inadequate formation of human capital and other nonenvironmental wealth in rural areas.

Must Economic Growth Always Harm the Environment?

Not very different from the simple Malthusian view is the opinion that economic growth inevitably damages natural resources. Pressure on the environment is bound to increase if living standards rise. In the final analysis, pessimists argue, that pressure cannot be accommodated.

If all this is true, Latin America's poor masses can only look forward to continued misery. For their lot to improve without an environmental cost, others—either the region's upper and middle classes or the citizens of wealthier nations—must get by with less. There is little reason to expect such a sacrifice to be made.

Fortunately, the relationship between economic development and resource use is much more flexible than what has been supposed by, say, those attempting to model environmental limits to growth (Meadows et al., 1972). At a macroeconomic level, adaptation to sharply higher oil prices during the 1970s shattered the widely shared belief that energy consumption is relatively insensitive to price changes and is instead determined by income levels (Sweeney, 1984). Of more direct relevance to this book, agricultural development, which increases commodity supplies, incomes, and employment, can be achieved in places like Ecuador without damaging renewable resources. Indeed, development and improved environmental quality can go hand in hand.

This latter possibility is illustrated by recent initiatives of the Brazilian Corporation for Agricultural Research (EMBRAPA). At EMBRAPA's National Soil Biology Research Center, scientists have isolated nitrogen-fixing bacteria that attach themselves to crops grown in hot and acidic soils. Most soybean farmers in the country now plant seeds inoculated with those bacteria. As a result, high soybean yields are achieved and annual expenditures on fertilizer, manufactured in part from nonrenewable resources, have been reduced by about $1 billion. EMBRAPA's biological pest control initiatives are also impressive. Scientists at its National Soybean Research Center have isolated a virus that kills the velvet bean caterpillar. Applying that virus to a soybean field costs 75 percent less than spraying pesticides does. In addition, toxic chemicals are not released into the environment (Mangurian, 1990).

The possibility of increasing crop and livestock production without encroachment on natural ecosystems is illustrated by recent Chilean experience. If yields had not risen in the country during the 1980s, annual growth in agricultural exports of 17.5 percent (Chapter 2) and annual population increases averaging 1.7 percent (IBRD, 1991b, p. 255) would have combined to cause cropland and pasture to increase by 1 percent a year or more (Southgate, 1991). But Chilean agriculture was simultaneously becoming more productive, in part because of increased mechanization, irrigation, and application of chemical inputs and also because use of improved varieties and cultivars was becoming more widespread (Arensberg et al., 1989). Since yields rose significantly, there was practically no agricultural land clearing between 1982 and 1987 (Southgate, 1991).

Is Economic Liberalization Inherently Unfair to Nature and the Poor?

Certainly in Latin America, few people have any difficulty understanding the basic lessons of recent agricultural development in Brazil and Chile. Official declarations issued by national governments and regional organizations forcefully articulate the position that improved use and management of renewable resources depend on economic progress and, conversely, that conservation is impossible where the economy is faltering and populations are poor and desperate (Comisión de Desarrollo y Medio Ambiente de América Latina y el Caribe, 1990). This view was reflected in the agenda for the June 1992 Earth Summit, held in Rio de Janeiro, where economic as well as environmental concerns were addressed.

Another consensus is emerging in Latin America, which is that development requires greater, not less, reliance on market forces. Chile, Mexico, and now Argentina have abandoned old models of import substitution and industrialization and, in some respects, are now among the world's most open economies. By regional standards, policy change in Ecuador has been tentative to date. But the right-of-center government that took power in August 1992 is slowly following the lead of Presidents Menem and Salinas, who have won impressive victories in mid-term elections after launching deep-rooted reforms.

Arguments that the public sector should closely regulate economic development are no longer credible. Critics of liberalization, then, fall back on two contentions. One is that increased reliance on market forces hurts the poor. The other is that liberalization damages the environment. If both these arguments were completely true, then the choices among growth, equity, and conservation facing every Latin American government would be vexing indeed.

The literature on the distributional consequences of economic reform in the region is still being written. One thing is clear, however. The macroeconomic and financial distortions that are a principal legacy of failed models of economic development are terribly unfair to the most vulnerable members of society. The Brazilian upper class, for example, has reaped sizable gains on financial markets because the country's government has been unable to control its deficits. Meanwhile, the poor majority has paid the price, in the form of high inflation. Any doubts about the inherent injustice resulting from interventionist policies are quickly dispelled by a review of recent Bolivian economic history. In the late 1980s, that country put an end to deficit spending, which had benefited a relatively affluent few and caused hyperinflation that ruined the many (Morales, 1988).

Now that most donor agencies, national governments, and economists are in favor of liberalization, the environmental movement has become a last redoubt for those who distrust market forces. Their position goes well beyond economists' concerns about externalities in an unregulated market economy (Baumol and Oates, 1988). For example, Osvaldo Hurtado, a respected former president of Ecuador and now a center-left pundit, stated flatly on May 28, 1992, that "economic liberalism is anti-ecological."

Environmental externalities are serious in Ecuador and effective policies for dealing with them must be developed immediately. The government's weak response to oil industry pollution in Amazonian rainforests (Chapter 7) and the damage that pesticide run-off seems to be doing to shrimp mariculture in the Guayas River Basin (Chapter 8) are representative of the challenge. Regardless, this book's analysis demonstrates conclusively that getting rid of policy-induced market distortions can often serve efficiency, equity, *and* environmental goals. The rural poor and the natural ecosystems they are obliged to mine suffer the consequences of financial sector repression and attenuated property rights in the countryside (Chapters 4 and 5). By contrast, some of the richest people in Ecuador have gained from subsidized irrigation water (Chapter 6), underpriced timber extracted from public forests (Chapter 4), cheap use permits for shrimp ponds located in coastal wetlands (Chapter 8), and nearly free licenses to operate cruise ships in the Galápagos (Chapter 9).

We have no doubt that affluent Latin Americans would absorb most of the costs of policy reform, which is a prerequisite for environmentally sound economic development.

Does Ecuador Have the Fiscal Capacity for Sustainable Development?

Allowing for the problem of externalities, existing production possibilities can be fully exploited if the forces of supply and demand are given free rein. To add to those possibilities, investment is needed. In this book, we have stressed the importance of human capital formation and improving the rural economy's scientific base. Investing in education can also have distributional payoffs. For example, the democratic government that succeeded the regime of General Pinochet sees human capital formation as the centerpiece of a strategy to spread the gains of market-based economic growth in Chile more evenly.

Accepting the argument that living standards and environmental quality can be improved through investment in nonenvironmental assets, many observers nevertheless remain pessimistic. It is nearly impossible, they say, to break out of poverty, and the resource degradation associated with it, because of adverse conditions beyond the control of countries like Ecuador. Declining terms of trade and the debt crisis are frequently cited as reasons why financing sustainable development is next to impossible.

Lower relative prices for foodstuffs and raw materials, which account for most exports from the developing world, can certainly inhibit economic growth and resource conservation. As was demonstrated by Ecuador's exertions to increase oil production during the 1980s in the face of falling energy prices, primary commodity producers often deplete resources more quickly in order to service debts and to maintain living standards. During that decade, the margin left over for investment was substantially reduced.

To be sure, agricultural protectionism in rich countries drives down commodity prices (FAO, 1992), to the detriment of producers in the developing world. For example, restrictions on banana imports into the European Economic Community, which went into effect in the middle of 1993, will cost Central and South America hundreds of millions of dollars in annual earnings and eliminate tens of thousands of jobs.

However, one must not forget that advances in production technology are also a major reason why raw materials and food are becoming cheaper, thereby presenting society as a whole with additional opportunities to do more than simply meet basic nutritional needs. Indeed, there is no surer sign of development, at either the international or national level, than a long-term decline in the prices of primary commodities brought about because of increasing productivity.

All things considered, the debt crisis of the 1980s has probably been a more serious impediment to human capital formation and other forms of investment in Latin America than falling prices for food and raw materials have been. However, the recent experience of several countries in the region demonstrates that the crisis is not insurmountable. For example, the Chilean and Mexican governments have diminished debt service burdens and renewed capital inflows by selling inefficient state enterprises and eliminating subsidies and the bureaucratic bloat that underlie chronic public sector deficits. That those two countries have largely put their debt problems behind them is indicated by the secondary market value of their bank loans. Chile's loans were trading for 91 percent of face value in May 1992. At the same time, Mexican discount bonds were being bought and sold for 82 cents on the dollar.

Ecuador is still mired in crisis, its old bank loans being traded at no more than 25 percent of face value on secondary markets in May 1992. However, the government could easily put its financial house in order. Consumer subsidies could be reduced. Parastatal enterprises, many of which hold valuable assets, could be sold. In addition, military spending could be trimmed.

Consumers are not as insulated from market forces as they used to be. In 1980, for example, fossil fuel subsidies peaked at 7 percent of GDP. The Borja administration deserves credit for closing much of the gap between domestic and border prices. After it assumed office in 1988, fuel subsidies were reduced to 2 percent of GDP (Thoumi, 1990, p. 46). However, the policy of bringing internal values into line with international ones has proven difficult to sustain. Because of inadequate adjustment after Iraq's invasion of Kuwait, for example, the ratio of domestic to border prices fell to 35 percent in September 1990. The World Bank has estimated that raising the ratio to 75 percent would eliminate Ecuador's public deficit (IBRD, 1991a, p. 44).

The prospect of higher energy prices always arouses strong opposition in Ecuador. Politicians tend to ignore the fact that a large share of the country's cheap fuel is smuggled to Peru and Colombia. Neither do they like to admit that subsidies are highly regressive. As is pointed out in Chapter 6, around 90 percent of the fuel and electricity sold in Quito and Guayaquil is consumed by the wealthiest 10 percent of those two cities' households (Kublank and Mora, 1987). The possibility that the general policy of cheap energy could be replaced with, say, public transportation subsidies, which are targeted more directly on the poor, is never raised. Instead, populist leaders merely bewail the suffering that, they say, all Ecuadorians are bound to endure if the gap between internal and border prices is closed.

Aside from reducing subsidies for rich consumers, the Ecuadorian government could finance productivity-enhancing investment by selling parastatal enterprises.

The country's wealthiest parastatal is Petroecuador, which has large oil reserves, refineries, pipelines, and other properties. Industry sources say that selling the company would bring in several billion dollars, which is credible in light of recent estimates of untapped oil deposits. Hicks et al. (1990) reported that proven reserves in 1988 were 1.1 billion barrels and that probable reserves amounted to 0.7 billion barrels. Deposits falling in the former category have increased since then because production has been exceeded by discoveries made by CONOCO, Atlantic Richfield Company, and other firms (Chapter 7). Petroecuador sources were reporting informally in May 1992 that the country has 2.3 billion barrels of untapped reserves. Multiplying Hicks et al.'s (1990) report of proven and probable reserves by a similarly conservative measure of average in situ values, $5 a barrel, yields a lower-bound estimate of what the company's natural resources are worth: $9.0 billion.

By selling Petroecuador, INECEL, the telephone company, an oil tanker firm, and other businesses, the national government could retire the national debt and interest arrears, which are approaching $13.0 billion. It would have enough money left over to set up a large development fund, which could be used to underwrite education, public health, family planning, and improvements in science and technology.

Other benefits would be realized as well. In 1988, consolidated operating losses of all state-owned businesses amounted to a third of the total nonfinancial public sector deficit, which was 5.1 percent of GDP (IBRD, 1991a, p. 23). Privatization would eliminate a large source of governmental red ink. The activities of a privatized petroleum industry and other companies formerly in government hands could also be taxed, thereby helping to shore up public finances. Furthermore, privatization should improve the quality of services. For example, electricity supplies and the telephone system would probably become more reliable and fuel quality could be expected to change for the better. Finally, it is hard to argue that natural resources would suffer. As Petroecuador's exemption from meaningful environmental regulation clearly demonstrates (Chapter 7), the government is unwilling to impose effective pollution controls on parastatal enterprises.

Finally, spending on the armed forces should be a prime target for budget cutters. The published defense budget for 1989 was 70 billion sucres (IISS, 1990, p. 194), equivalent to $132 million at the BCE intervention exchange rate of 526.35 sucres to the dollar. In addition, the military received 29.5 percent of the value of oil exports (which equaled $1.02 billion). Total allocations from the government, then, were $437 million, or a little less than 5 percent of GDP in 1989. There are also the retained earnings of various firms owned by the armed forces, which have grown to be the largest business group in the country (in part because military firms are shielded in various ways from foreign and domestic competition). No accurate figures are available, largely because financial reports of enterprises owned entirely by the military are kept strictly confidential. However, some observers, including some of the government's own economists, claim that the sum of governmental allocations and retained earnings might approach 10 percent of GDP.

Without a doubt, reducing consumer subsidies, selling parastatals, and cutting military spending all require that highly influential groups give ground. However, the advantage of macroeconomic stability, which will come once fiscal deficits have been brought decisively under control, is an important one. Furthermore, the long-term benefits of eliminating distorted public spending should be enormous. Had the billions of dollars that Brazil, for example, channeled over the years to industries that create relatively little employment and are too inefficient to compete in international markets been directed instead to education and improved public health services, it would be a very different nation today. Rather than being saddled with millions of people fit only for menial work, like agricultural land clearing, Brazil's economy would be much larger and its natural resources would be in much better shape.

The scale of human and environmental dislocation resulting from erroneous development strategies is certainly greater in Brazil than it is in Ecuador. But in the latter country as well as the former, the same question needs to be faced squarely. Where half the citizenry is plagued by illiteracy and disease and where strengthening of the rural economy's scientific base is needed to reduce renewable resource mining in the countryside, why does the government choose to hold billions of dollars worth of oil deposits, petroleum refineries, tankers, dams, transmission lines, grain elevators, airplanes, and on and on? The inefficiency and inequity of the status quo should be immediately obvious to anyone.

11

Resolving the Policy Crisis

As developed and practiced in North America, Europe, and other affluent parts of the world, natural resource economics focuses mainly on market failure. A typical subject of analysis is the inefficiencies that arise when a competitive and unregulated group of firms or households fails to internalize all the environmental costs of production or consumption. Analysis of externality problems often leads to recommendations for remedial intervention by the government, in the form of effluent taxes, emissions guidelines, and the like.

As is acknowledged at the beginning of this book, market failure is pervasive in the developing world. The inhabitants of large African, Asian, and Latin American cities are choking on air pollution, which has grown excessive because people who drive vehicles or run factories consider the environmental consequences of their actions to be externalities. At a global level, few nations consider humankind's interests in climatic stability and biological diversity when making decisions about industrial development and agricultural land clearing.

In any part of the world, implementing the policies needed to reduce local environmental externalities, like pollution of some river, lake, or stream, is often a challenging task. By the same token, it can be difficult to reconcile Africans', Asians', and Latin Americans' desire for economic growth with a commonly shared concern for global environmental quality. Where poor countries must forego material gain in order, say, to limit the greenhouse effect, compensation is warranted. How to effect international transfers is, of course, an important problem facing policy makers as well as economists.

Notwithstanding the importance of externality problems in places like Ecuador, this book does not focus primarily on market failure. That is, we have not concentrated on describing ways that the public sector can interfere with firms' and households' choices so that environmental costs will be internalized. Instead, our principal concern has been government failure, the central thesis being that current national policies discourage the wise use and management of natural resources in rural areas by impeding and distorting economic development.

As is made clear in the preceding chapter, economic progress and environmental conservation can be complementary and are also well within Ecuador's reach. To achieve sustainable development, comprehensive domestic policy reform is required. The process of changing macroeconomic and sectoral policies that discriminate against economic activity in the countryside has to be pursued with vigor. Laws and regulations that distort markets for natural resources, labor, and other inputs must likewise be eliminated. Furthermore, the government can concentrate more on facilitating investment in human capital and the scientific base as it stops interfering with market forces and private property rights. That interference frequently involves large budgetary outlays; therefore, the funds required to control environmental externalities, like oil industry pollution in the Oriente and agricultural chemical run-off from farms, can only become available if economic liberalization proceeds.

To conclude this chapter's prescriptive discussion, we acknowledge that integrated reform is a challenge. Piecemeal changes in the general policy environment are likely to be futile and could even be counterproductive. In addition, at least some specific proposals will arouse strong opposition. But unless the reforms outlined in the pages that follow are carried out fully, places like Ecuador can expect to endure unrelenting poverty and accelerated resource degradation for some time to come.

Nondiscriminatory Macroeconomic and Sectoral Policies

As is the case throughout Latin America (Krueger, Schiff, and Valdés, 1988), Ecuador embraced a policy environment that was strongly biased against agriculture and other sectors of the rural economy for many years. Because the sucre was chronically overvalued, domestic producers found it difficult to compete with imports and were discouraged from shipping goods to foreign markets. Bowing to urban and industrial interests, the national government kept prices for crops, livestock, timber, and other commodities low, through direct controls and export restrictions, as well as currency distortions. In terms of forgone GDP, Ecuador has paid a very high price for this biased regime (Scobie, Jardine, and Greene, 1990).

In 1981, elimination of discriminatory macroeconomic and sectoral policies began. Having proceeded at an uneven pace over the years, reform accelerated in August 1992, when a new administration took office.

Some of the environmental impacts of policy change could be negative. Stimulating economic activity in the countryside will probably cause some renewable resources that formerly sat idle to be brought into production. For example, an increase in crop and livestock prices could lead to increased encroachment by farmers and ranchers on tropical forests and other natural habitats.

However, the environmental benefits of a policy regime that is free of intersectoral biases cannot be ignored. A recently completed evaluation of the impacts on natural resource management of structural adjustment in the Philippines showed that import substitution and industrialization policies have encouraged people to deplete natural resources in the rural economy in order to invest in other sectors (Cruz and Repetto, 1992). The same thing has happened in Ecuador, where large positive trade balances in agriculture have been used to subsidize urban-based industry over the years (Whitaker and Greene, 1990).

The relative strength of different impacts of macroeconomic and sectoral policy reform is an interesting topic for future empirical research. For now, though, it is possible to say only that removing distortions in markets for outputs is a necessary, though not sufficient, condition for the wise use and management of the natural environment. For increasing resource scarcity to be accommodated, policy-induced distortions in factor markets must be eliminated simultaneously.

Efficient Markets for Natural Resources and Other Inputs

As is indicated in Chapter 3 and in several of this book's case studies, markets for natural resources and other inputs are prevented from performing their most important function, which is to transmit scarcity values. For this function to be reinforced, the public sector has to relinquish control over resources. In addition, the mechanisms employed to transfer properties to private hands must be structured to encourage conservation.

Our analysis also shows that improved performance of real estate and financial markets, across the board, is essential if firms and households are to respond efficiently to mounting renewable resource scarcity. An investment in "institutional infrastructure" (e.g., modern land registries) and elimination of legal restrictions on private tenure are needed if real estate markets are to work better. Now that laws and regulations that keep real interest rates negative are largely a thing of the past, improved financial intermediation depends primarily on eliminating inflation.

Denationalizing Natural Resources

We have stressed in this volume that there is a great imbalance between the public sector's extensive claims on the environment and its limited capacity to manage and to control access. The government is not taking very good care of Ecuador's forests (Chapter 4), water (Chapter 6), coastal ecosystems (Chapter 8), and unique habitats (Chapter 9). Damage to the environment has been compounded when the government has charged too little for the properties it sells or has made the removal of natural ecosystems a condition for private resource rights.

Fortunately, Ecuador, like other Latin American countries, is moving away from tenurial arrangements typical of agricultural frontiers. The law no longer obliges colonists to clear away trees in order to win an adjudication from IERAC. But the tradition of agricultural use rights, which has a long history in the Western Hemisphere, is fading slowly in many rural areas (Chapter 4).

The government is also raising the prices at which it sells resources to the private sector. Water tariffs have increased substantially in 1993. At the same time, the fee paid by loggers extracting timber from the Forest Patrimony has been raised from 1,350 to 4,000 sucres ($0.71 to $2.11) per cubic meter. In addition, the owners of Galápagos cruise ships must now pay INEFAN $200 a year for each passenger berth, which is substantially higher than the former charge (Chapter 9).

In spite of this progress, subsidies for the use of public resources are still high. Because a wide gap remains between the price and cost of irrigation water, existing systems continue to be inefficient and there is still political pressure to invest in expensive new projects that cannot be justified on economic grounds. In addition, paying a pittance for the use permit needed to build a shrimp pond below the high-tide line is a major reason why coastal ecosystems are being destroyed (Chapter 8).

There is no simple recipe for getting resource tenure and prices right. For example, agricultural water subsidies cannot be eliminated overnight. Since those subsidies have been capitalized into real estate values (Chapter 6), bringing prices into line with costs immediately would put financial stress on people who have taken out loans to buy irrigated land. Some would go bankrupt. The solution is to phase out subsidies and to spend more on research and extension (e.g., the transfer of improved water management techniques to farmers). The latter measure enhances the ability of irrigation systems' beneficiaries to pay higher prices. Of course, marginal-cost pricing should be the policy governing all future agricultural water development.

As a way to transfer properties to firms and individuals, competitive auctions hold some appeal. Provided that the external impacts of resource use and management are not very great, winning bids tend to reflect resource values. For equity's sake, people living close to auctioned resources can be given preferential treatment. One way to do this is to multiply prices offered by local individuals and communities by some factor greater than one.

There are some opportunities to use resource auctions in Ecuador. Private companies could bid for the right to take tourists to the Galápagos and other national parks. Public lands that are good sites for shrimp ponds could be sold the same way. For auctions to be truly competitive, participation by foreigners as well as Ecuadorians would have to be allowed.

As in many other developing countries, the best approach to resource denationalization in Ecuador is simply to recognize the de facto rights of local individuals and groups. Practically all of the Forest Patrimony, for example, is already occupied. In many cases, forest-dwelling communities established themselves long before boundaries for national parks and other government holdings were demarcated. It is only fair for their long-standing land rights to be recognized fully.

Letting associations of water users take over irrigation systems is proving to be a good alternative to public sector management. The case of the 8,000-hectare Azua irrigation project in the Dominican Republic is instructive. Farmers there have responded to the opportunity to run secondary and tertiary canals by increasing operation and maintenance expenditures fivefold. Water use has been reduced by 40 percent and farm incomes have doubled (Willardson and Anderson, 1993).

There is a precedent in Ecuador for giving local communities responsibility for the administration of public irrigation systems. Nearly ten years ago, INERHI turned over the Patate project, in the central Sierra, to local farmers. This change having been well received, two other small projects in the same region will be transferred to local communities in the near future. CEDEGE is seriously investigating options to public sector management of the irrigation system put in during the first phase of the Daule-Peripa Project (Table 6-1). In particular, there is a proposal to sign a contract with a private firm, which would provide management and training services during a transition period that would end with control of the system being passed to a water users' association.

Regardless of how they are distributed, property rights in denationalized resources must be structured carefully. A timber concession, for instance, should last for a few decades at least. Otherwise, the typical concession holder will not want to replant or to manage secondary regeneration, either of which increases future harvests. In addition, resource privatization will not solve all environmental problems. For example, owners are unlikely to leave mature forests untouched for very long since timber growth rates in those forests are generally less than rates of return for alternative investments.

However, the merits of privatization as an alternative to governmental management are undeniable. As experience with irrigation systems in the Dominican Republic and other countries clearly demonstrates, giving secure property rights to water users can lead to increased production and income as well as improved resource management.

The distributional consequences would be positive as well. As is indicated in Chapter 6, relatively few farmers benefit from the large sums required to keep irrigation water cheap in existing systems and to build new dams and canals; many of those farmers are relatively affluent. Limited budgets for agricultural development could instead be used for research and for technology transfer, which create sector-wide impacts. At a macroeconomic level, reducing irrigation subsidies would help to alleviate the public sector deficits that are the root cause of inflation, which is an especially severe affliction for poor people in countries like Ecuador.

Eliminating Policy-Induced Inefficiencies in Real Estate and Financial Markets

Ecuadorian factor markets are not out of kilter solely because of the problems resulting from excessive public sector claims on natural resources. Throughout the country, rural land and financial markets function poorly, both because private property rights are weak and because the price mechanism is subject to government interference.

The practical effect of attenuated property rights is to load down real estate markets with transactions costs. This, in turn, discourages adoption of conservation measures because land owners know they cannot capture the full value of those and other improvements when they sell their holdings.

Transactions costs can be reduced by changing land reform laws and the Commune Law. A fundamental goal of agrarian reform legislation passed in 1964 and amended in subsequent years was to eliminate large underutilized estates. Since recent research shows that few of those estates still exist (Camacho and Navas, 1991), it does not make sense to continue IERAC's costly overview of the sale, rental, and use of agricultural land. Neither can the threat of expropriation, which current land reform laws maintain, be justified. By the same token, there is no rationale for the prohibition on dissolving collective holdings, which keeps commune members from participating in land and financial markets. Ecuador needs to follow the lead of Latin American countries such as Mexico, where President Salinas has declared his intention "to end the possibility of land reform" (Fraser, 1991).

While important, eliminating agrarian reform legislation and giving communes the right to break themselves up will not be enough to make real estate markets function efficiently in rural Ecuador. As we point out in Chapter 3, land registries in the countryside are an antiquated shambles. Through pilot projects conducted in two local jurisdictions from the late 1980s through 1991, AID has developed a cost-effective system for demarcating and recording property rights (Lambert et al., 1990; Moreno, 1992). The investment required to apply this system throughout Ecuador would yield large dividends in the form of facilitated real estate transactions.

Improving the efficiency of rural land markets would greatly improve financial intermediation in the countryside. Making use of workable land registries, lenders would have a much easier time evaluating loans backed by real estate. The value of that collateral would be further enhanced if agrarian reform laws were eliminated, thereby making land transactions easier and removing expropriation risks.

In the past, financial markets have functioned poorly because interest rates were not determined by supply and demand (Chapter 3). Laws permitting the application of controls still exist, although the chances are small that they will be reintroduced in the near future. However, making loans and saving continue to be discouraged in Ecuador by inflation. Largely because of the uncertainties that inflation creates, credit has grown particularly scarce in the countryside.

If outdated agrarian reform laws can be changed, modern land registries built up, inflation reduced, and interest rates kept unregulated, real estate and financial markets should function efficiently in rural areas. All of this will put producers of crops, livestock, and other commodities in a better position to make the investments required to use more manufactured inputs instead of natural resources. They will also be more inclined to adopt conservation measures and other land improvements.

Formation of Human Capital and Strengthening the Rural Economy's Scientific Base

Over the years, the Ecuadorian government has paid heavily for its interference with market forces. At the best of times, regressive energy subsidies have eaten up 2 or 3 percent of national GDP. Within the rural economy, large sums have been spent on irrigation subsidies and on progressively larger and more wasteful new projects. As the World Bank found in its recently completed survey of public finance (IBRD, 1991a), there are dozens of examples of wasteful spending in the public sector.

Relative to the billions of dollars of assets that the government currently holds (Chapter 10), the investment required to improve agricultural productivity is negligible. International organizations recommend that 1 to 2 percent of the value of a country's crop and livestock output be allocated to research in support of the sector. To conform to this standard, Ecuador would have to spend only $15 to $30 million. Much of the required amount could be raised from the private sector. One way to do this is with a checkoff scheme (i.e., a self-imposed tax agreed to by a group of private producers).

Recent Brazilian and Chilean experience (Chapter 10) demonstrates the environmental benefits that Ecuador can enjoy as a result of this minor investment. Ways to grow crops with less chemical fertilizers and pesticides can be found and applied. In addition, increasing demand for agricultural commodities can be satisfied with less encroachment on natural ecosystems.

Much more money is needed to broaden access to high-quality education. The precise rate at which the earnings generated by natural resource exploitation should be channeled to the formation of nonenvironmental wealth is easier to characterize theoretically (Hartwick, 1977) than to define empirically. Nevertheless, it is clear that no factor will shape Ecuador's economic and environmental future more than human capital formation. The clear lesson of development in the late twentieth century is that nothing is more important than human resources. At the individual level, investing in education allows people to move into more remunerative forms of employment. To compete internationally, any country must have a well-trained labor force.

The environmental consequences of inadequate human capital formation are abundantly clear throughout Latin America. Along Andean hillsides, in Amazonian jungles, and elsewhere, individuals and families without the skills to land better jobs (because funds that could have been directed to human capital formation were dissipated instead on subsidies for consumption and inefficient industries) must eke out a marginal living by mining soils, forests, and other renewable resources. Renewable resource degradation is in large part a visible and nonsustainable manifestation of the rural poor's despair, brought about by misguided economic policies.

The Challenge of Reform

The difficulty of seeing through effective policy change should not be underestimated. The status quo receives strong support from influential quarters. Industrial and other urban interests, for example, have benefited from macroeconomic and sectoral policies that are biased against the rural economy. Also, our general recommendation to continue raising resource prices is bound to generate opposition. Farmers will not want to pay more for irrigation water and loggers will resist higher stumpage fees, just as some tourism companies were reluctant to pay more for continued access to the Galápagos. Furthermore, many employees of governmental agencies and parastatals that are slated to disappear or to be transferred to the private sector are bound to be distressed. The military's traditionally strong resistance to even minor budget cuts is sure to continue.

Reform is also a challenge because piecemeal changes in policy are likely to be ineffective. Development and dissemination of, say, better forest management techniques and tree species that grow quickly will have modest results if access to public forests remains essentially free. Conversely, strengthening property rights in tree-covered land will not reduce deforestation greatly if geographic expansion of the agricultural economy remains the principal response to increasing commodity demands. This will be the case if spending on agricultural research and extension is not increased.

Partial reform can even be counterproductive. In particular, removing policies that keep prices for crops, timber, and other commodities low could accelerate encroachment on natural ecosystems, if the open-access status of those resources is not changed.

However, we are optimistic. The Ecuadorian government is no longer able to pay for subsidies and programs that serve no good purpose. We look forward to a time when macroeconomic and sectoral policy distortions are eliminated, factor markets are allowed to function efficiently, and the public sector, freed of the financial burdens inherent in old policies, concentrates instead on facilitating productivity-enhancing investment. That time will be marked by rapid economic growth as well as sound natural resource management.

Abbreviations

AID	U.S. Agency for International Development
BCE	Banco Central del Ecuador (Central Bank of Ecuador)
BNF	Banco Nacional de Formento (National Development Bank)
CEDEGE	Comisión de Estudios para el Desarrollo de la Cuenca del Rio Guayas (Commission for the Study of the Development of the Guayas River Basin)
CELADE	Centro Latinoamericano de Demografía (Latin American Demographic Center)
CEPAR	Centro de Estudios de Población y Paternidad Responsable (Center for the Study of Population and Responsible Parenthood)
CEPE	Corporación Estatal Petrolera Ecuatoriana (Ecuadorian Oil Company)
CIP	Centro Internacional de la Papa (International Potato Center)
CLIRSEN	Centro de Levantamientos Integrados de Recursos Naturales por Sensores Remotos (Center for the Integrated Survey of Natural Resources through Remote Sensing)

CONACYT	Consejo Nacional de Ciencia y Tecnología (National Council on Science and Technology)
CONADE	Consejo Nacional de Desarrollo (National Development Council)
CPC	Cámara de Productores de Camarón (Chamber of Shrimp Producers)
CREA	Centro de Reconversión Económico del Azuay, Cañar, y Morona Santiago (Center for the Economic Recovery of Azuay, Cañar, and Morona Santiago)
CRM	Centro de Rehabilitación de Manabí (Center for the Rehabilitation of Manabí)
DGMM	Dirección General de la Marina Mercante (Merchant Marine Directorate)
DGP	Dirección General de Pesca (General Fishing Directorate)
DIGEMA	Dirección General de Medio Ambiente (General Directorate of the Environment)
DINAMA	Dirección Nacional de Medio Ambiente (National Directorate of the Environment)
EMAP-Q	Empresa Municipal de Agua Potable de Quito (Quito Water Company)
EMBRAPA	Empresa Brasileira de Pesquisa Agrícola (Brazilian Corporation for Agricultural Research)
EPA	U.S. Environmental Protection Agency
FAO	Food and Agriculture Organization of the United Nations
FONARYD	Fondo Nacional para Riego y Drenaje (National Fund for Irrigation and Drainage)
IBRD	International Bank for Reconstruction and Development (World Bank)
IDB	Inter-American Development Bank
IDEA	Instituto de Estrategias Agropecuarias (Agricultural Strategies Institute)
IERAC	Instituto Ecuatoriano de Reforma Agraria y Colonización (Ecuadorian Institute for Agrarian Reform and Colonization)
IICA	Instituto Interamericano de Cooperación para la Agricultura (Intcramerican Institute for Agricultural Cooperation)

IISS	International Institute for Strategic Studies
INEC	Instituto Nacional de Estadísticas y Censos (National Institute of Statistics and Censuses)
INECEL	Instituto Ecuatoriano de Electrificación (Ecuadorian Institute for Electrification)
INEFAN	Instituto Ecuatoriano Forestal, de Areas Naturales, y de Vida Silvestre (Ecuadorian Institute of Forestry, Natural Areas, and Wildlife)
INERHI	Instituto Ecuatoriano de Recursos Hidráulicos (Ecuadorian Institute of Water Resources)
INIAP	Instituto Nacional de Investigaciones Agropecuarias (National Institute of Agricultural Research)
IUCN	International Union for the Conservation of Nature
MAG	Ministerio de Agricultura y Ganadería (Ministry of Agriculture and Livestock)
MEM	Ministerio de Energía y Minas (Ministry of Energy and Mines)
PREDESUR	Programa Regional para el Desarrollo del Sur (Regional Program for the Development of Southern Ecuador)
PRONAMEC	Programa Nacional de Mecanización (National Mechanization Program)
PRONAREG	Programa Nacional de Regionalización Agraria (National Program for Agrarian Regionalization)
RAN	Rainforest Action Network
SEAN-INEC	Subsecretaría Forestal y de Recursos Naturales del Instituto Nacional de Estadísticas y Censos (National Agricultural Statistics System of the National Institute of Statistics and Censuses)
SMA	Subsecretaría de Medio Ambiente (Subsecretariat of the Environment)
SPNG	Servicio del Parque Nacional Galápagos (Galápagos National Park Service)
SUFOREN	Subsecretaría Forestal y de Recursos Naturales (Subsecretariat for Forestry and Natural Resources)
UNESCO	United Nations Educational, Scientific, and Cultural Organization
WCED	World Commission on Environment and Development
WRI	World Resources Institute

References

Arensberg, W., M. Higgins, R. Asenjo, F. Ortiz, and H. Clark. 1989. "Environment and Natural Resources Strategies in Chile," U.S. Agency for International Development, Santiago.
Arriaga, L. 1989. "The Daule-Peripa Dam Project, Urban Development of Guayaquil and Their Impact on Shrimp Mariculture" in S. Olsen and L. Arriaga (eds.), *A Sustainable Shrimp Industry for Ecuador*. Narragansett: University of Rhode Island Coastal Resources Center.
Banco Central del Ecuador (BCE). 1989. *Memoria Anual: 1988*. Quito.
Banco Central del Ecuador (BCE). 1990. "La Actividad Petrolera en el Ecuador en la Década de los 80," División Técnica, Quito.
Banco Central del Ecuador (BCE). 1991. *Boletín Anuario No. 13: 1991*. Quito.
Baumol, W., and W. Oates. 1988. *The Theory of Environmental Policy*. Cambridge: Cambridge University Press.
Blaikie, P. 1985. *The Political Economy of Soil Erosion in Developing Countries*. London: Longman.
Bromley, R. 1981. "The Colonization of Humid Tropical Areas in Ecuador," *Singapore Journal of Tropical Geography* 2:1, pp. 15–26.
Bulatao, R., E. Bos, P. Stephens, and M. Vu. 1990. *World Population Projections 1989–90 Edition*. Baltimore: Johns Hopkins University Press.
Camacho, C., and R. Flores. 1993. "El Rol del Banco Nacional de Fomento en el Sistema de Intermediación Financiera" (report to Banco Nacional de Fomento and World Bank), Instituto de Estrategias Agropecuarias, Quito.
Camacho, C., and M. Navas. 1991. "Evaluación del Processo de Cambio en la Tenencia de la Tierra en la Sierra Norte y Central" (Technical Document No. 41), Quito.

Cámara de Productores de Camarón (CPC). 1989. "Libro Blanco del Camarón," Ecuagraf, Guayaquil.
Caujolle-Gazet, A., and C. Luzuriaga. 1986. "Estudio de un Tipo de Cangahua en el Ecuador: Posibilidades de Mejoramiento Mediante el Cultivo" in *La Erosion en el Ecuador*. Quito: Centro de Investigación Geográfica.
Centro de Estudios de Población y Paternidad Responsable (CEPAR). 1990. *Ecuador: Encuesta Demográfica y de Salud Materna e Infantil, 1989*. Quito.
Colyer, D. 1990. "Agriculture and the Public Sector" in M. Whitaker and D. Colyer (eds.), *Agriculture and Economic Survival: The Role of Agriculture in Ecuador's Development*. Boulder: Westview Press.
Comisión de Desarrollo y Medio Ambiente de América Latina y el Caribe. 1990. *Nuestra Propia Agenda*. Washington: Inter-American Development Bank.
Commander, S., and P. Peek. 1986. "Oil Exports, Agrarian Change, and the Rural Labor Process: The Ecuadorian Sierra in the 1970s," *World Development* 14:1, pp. 79–96.
Consejo Nacional de Desarrollo, Instituto Nacional de Estadística y Censos, and Centro Latinamericano de Demografía (CONADE/INEC/CELADE). 1993. *Ecuador: Estimaciones y Proyecciones de Población, 1950–2010*. Quito: Instituto Nacional de Estadística y Censos.
Corporación Estatal Petrolera Ecuatoriana (CEPE). 1987. "Análisis de la Contaminación Ambiental en los Campos Petroleros Libertador y Bermejo," Quito.
Crissman, C., and D. Cole. 1994. "Pesticide Use and Farm Worker Health in Ecuadorian Potato Production," Allied Social Science Associations Meetings, Boston.
Crissman, C., and P. Espinosa. 1992. "Agricultural Chemical Use and Sustainability of Andean Potato Production: Project Design and Pesticide Use in Potato Production in Ecuador," Rockefeller Foundation Seminar on Measuring the Health and Environmental Effects of Pesticides, Bellagio.
Cruz, R., A. Orquera, and A. Salazar. 1986. "El Riego en el Ecuador," Instituto Ecuatoriano de Recursos Hidráulicos, Quito.
Cruz, W., and R. Repetto. 1992. *The Environmental Effects of Stabilization and Structural Adjustment Programs: The Philippines Case*. Washington: World Resources Institute.
Darwin, C. 1845. *The Voyage of the Beagle*. London: John Murray.
Darwin, C. 1859. *On the Origin of the Species by Means of Natural Selection*. London: John Murray.
de Groot, R. 1983. "Tourism and Conservation in the Galápagos Islands," *Biological Conservation* 26:4, pp. 291–300.
Delavaud, A. 1982. *Atlas del Ecuador*. Paris: Les Editions.
de Noni, G., and G. Trujillo. 1986. "La Erosión Actual y Potencial en Ecuador: Localización, Manifestaciones, y Causas" in *La Erosión en el Ecuador*. Quito: Centro de Investigación Geográfica.
Detwiler, R., and C. Hall. 1988. "Tropical Forests and the Global Carbon Cycle," *Science* 239, pp. 42–47.
Dodson, C., and A. Gentry. 1991. "Biological Extinction in Western Ecuador," *Annals of the Missouri Botanical Garden* 78:2, pp. 273–295.
Edwards, S. 1991. "The Demand for Galápagos Vacations: Estimation and Application to Conservation," *Coastal Management* 19:2, pp. 155–169.

Empresa Municipal de Agua Potable de Quito (EMAP-Q). 1991. "Tabla de Facturación: Consumo Doméstico," Quito.

Estrada, R. 1993. "Incidencia de las Políticas Económicas en la Conservación de los Recursos de la Zona Andina," Reunión de Planificación y Priorización de Actividades de Investigación sobre el Agroecosistema Andino, International Potato Center, Lima.

Fearnside, P. 1987. "Summary of Progress in Quantifying the Potential Contribution of Amazonian Deforestation to the Global Carbon Cycle" in D. Athie, T. Lovejoy, and P. Owens (eds.), *Proceedings of the Workshop on Biogeochemistry of Tropical Rainforests: Problems for Research*. Piracicaba: Universidade de Sao Paulo Centro de Energía Nuclear na Agricultura.

Food and Agriculture Organization of the United Nations (FAO). 1989. *Production Yearbook, Volume 42 (1988)*. Rome.

Food and Agriculture Organization of the United Nations (FAO). 1992. *The State of Food and Agriculture, 1991*. Rome.

Fraser, D. 1991. "Land Reform: A Second Agricultural Revolution Under Way," *Financial Times*, October 25, p. VI.

Gonzalez-Vega, C. 1984. "Credit Rationing Behavior of Agricultural Lenders" in D. Adams, D. Graham, and J. von Pischke (eds.), *Undermining Rural Development with Cheap Credit*. Boulder: Westview Press.

Grimes, A., S. Loomis, P. Jahnige, M. Burnham, K. Onthank, R. Alarcón, W. Palacios, C. Carón, D. Niell, M. Balick, B. Bennett, and R. Mendelsohn. 1993. "Valuing the Rain Forest: The Economic Value of Non-Timber Forest Products in Ecuador," Yale University School of Forestry, New Haven.

Hachette, D., and D. Franklin. 1990. "Employment and Incomes in Ecuador: A Macroeconomic Context," U.S. Agency for International Development, Quito.

Harden, C. 1991. "Andean Soil Erosion," *National Geographic Research and Exploration* 7:2, pp. 216–231.

Hardin, G. 1968. "The Tragedy of the Commons," *Science* 168, pp. 1243–1248.

Hartwick, J. 1977. "Intergenerational Equity and the Investing of Rents from Exhaustible Resources," *American Economic Review* 67:5, pp. 972–974.

Hayami, Y., and V. Ruttan. 1985. *Agricultural Development: An International Perspective* (2nd ed.). Baltimore: Johns Hopkins University Press.

Hendrix, S. 1993. "Latin American Property Law Modernization," Thirtieth Conference of the Committee on Commercial Law and Procedure of the Inter-American Bar Association, Santiago.

Hicks, J., H. Daly, S. Davis, and M. de Freitas. 1990. "Ecuador's Amazon Region: Development Issues and Options" (Discussion Paper No. 75), International Bank for Reconstruction and Development, Washington.

Instituto Ecuatoriano de Electrificación (INECEL). 1988. *Reportes Anuales de Operación de INECEL: 1988*. Quito.

Instituto Ecuatoriano de Electrificación (INECEL). 1988–1992. *Estado Financiero de INECEL*. Quito: Departamento de Operaciones del Sistema Nacional Interconectado.

Instituto Ecuatoriano de Electrificación (INECEL). 1992. "Precios Medios a Nivel Nacional de las Empresas Eléctricas," Dirección de Estudios y Control de Tarifas, Quito.

Instituto Ecuatoriano de Recursos Hidráulicos (INERHI). 1987. "Tarifa de Riego, 1988," Quito.
Instituto Interamericano de Cooperación para la Agricultura (IICA). 1988. *Ajuste Macroeconómico y Sector Agropecuario en América Latina.* Buenos Aires.
Instituto Nacional de Estadística y Censos (INEC). 1984. *Encuesta Nacional de Fecundidad, 1979.* Quito.
Instituto Nacional de Estadística y Censos (INEC). 1991. *Resultados Definitivos del V Censo de Población y IV de Vivienda, Resumen Nacional.* Quito.
Inter-American Development Bank (IDB). 1981. "Expert Evaluation of the Montúfar Irrigation Project," Quito.
International Bank for Reconstruction and Development (IBRD). 1982. "Informe de Evaluación Ex Post, Ecuador: Proyecto de Riego Milagro," Department of Evaluation of Operations, Washington.
International Bank for Reconstruction and Development (IBRD). 1991a. *Ecuador: Public Sector Reforms for Growth in the Era of Declining Oil Ouput.* Washington.
International Bank for Reconstruction and Development (IBRD). 1991b. *World Development Report 1991.* Washington.
International Institute for Strategic Studies (IISS). 1990. *The Military Balance, 1990–1991.* Oxford: Oxford Nuffield Press.
Jackson, M. 1990. *Galápagos: A Natural History Guide.* Calgary: University of Calgary Press.
Keller, J., A. La Baron, R. Meyer, D. Daines, and D. Anderson. 1982. "Ecuador: Irrigation Sector Review" (Report No. 12), U.S. Agency for International Development Water Management Synthesis Project, Utah State University, Logan.
Kimerling, J. 1991a. *Amazon Crude.* Washington: Natural Resources Defense Council.
Kimerling, J. 1991b. "Disregarding Environmental Law: Petroleum Development in Protected Natural Areas and Indigenous Homelands in the Ecuadorian Amazon," *Hastings International and Comparative Law Review* 14:4, pp. 849–903.
Krueger, A., M. Schiff, and A. Valdés. 1988. "Agricultural Incentives in Developing Countries: Measuring the Effect of Sectoral and Economy-Wide Policies," *World Bank Economic Review* 2:1, pp. 255–271.
Kublank, P., and D. Mora. 1987. "El Sistema Energético del Ecuador," Instituto Latinoamericano de Investigaciones Socioeconómicas, Quito.
Lambert, V., A. Frank, A. Cisneros, D. Sherrill, B. DeWalt, A. Abellan, and J. Uquillas. 1990. "Ecuador Land Titling Project Evaluation," U.S. Agency for International Development, Quito.
Landázuri, H., and C. Jijón. 1988. *El Medio Ambiente en el Ecuador.* Quito: Instituto Latinoamericano de Investigaciones Socioeconómicas.
LiPuma, E., and S. Meltzoff. 1985. "The Social Economy of Shrimp Mariculture in Ecuador," Rosenstiel School of Marine and Atmospheric Science, University of Miami, Miami.
Machlis, G., D. Costa, and J. Cárdenas-Salazar. 1990. "Estudio del Visitante a las Islas Galápagos," Fundación Charles Darwin, Quito.
Mahar, D. 1989. "Government Policies and Deforestation in Brazil's Amazon

Region," International Bank for Reconstruction and Development, Washington.
Mangurian, D. 1990. "Enlisting Nature's Aid," *IDB Newsletter*, September–October, p. 3.
Marshall-Silva, J. 1988. "Ecuador: Windfalls of a New Exporter" in A. Gelb (ed.), *Oil Windfalls: Blessing or Curse*. Oxford: Oxford University Press.
McPadden, C. 1985. "A Brief Review of the Ecuadorian Shrimp Industry," Instituto Nacional de Pesca, Guayaquil.
Meadows, D., D. Meadows, J. Randers, and W. Behrens. 1972. *The Limits to Growth*. New York: Universe Books.
Ministerio de Agricultura y Ganadería (MAG). 1987. "Informe Final de la Limitación del Patrimonio Forestal del Estado," Quito.
Ministerio de Agricultura y Ganadería and Consejo Nacional de Ciencia y Tecnología (MAG/CONACYT). 1986. "Proyecto de Determinación de Residuos de Pesticidas Clorados en Leche Materna," Quito.
Montenegro, F. 1987. "Plantaciones Forestales Productivas en los Trópicos del Ecuador con Pachaco" and "Plantaciones Forestales Productivas en los Trópicos del Ecuador con Laurel" in I. McCormick (ed.), *Análisis Económico de Inversiones en Plantaciones Forestales en Ecuador*. Quito: Asociación de Industrias Madereras.
Morales, J. 1988. "Inflation Stabilization in Bolivia" in M. Bruno, G. Di Tella, R. Dornbusch, S. Fischer, E. Helpman, N. Liviatan, and L. Meridor (eds.), *Inflation Stabilization*. Cambridge: Massachusetts Institute of Technology Press.
Moreno, R. 1992. "Ecuador's Rural Cadasters and Land Titling Project (CATIR): Technical Processes," *Journal of Surveying Engineering* 118:4, pp. 118–229.
Mountfort, G. 1974. "The Need for Partnership: Tourism and Conservation," *Development Forum* 2:3, pp. 6–7.
Moya, R., and N. Peñafiel. 1989. "Estimación de la Pérdida de Suelo Producida por Erosión Hídrica en el Cultivo de Maíz, Cangahua, Pichincha," Facultad de Ciencias Agrícolas de la Universidad Central, Quito.
Município de Quito. 1992. "Costos de Capital, Operaciones y Mantenemiento: Papallacta," Unidad Ejecutora del Proyecto Papallacta, Quito.
Myers, N. 1988. "Threatened Biotas: Hotspots in Tropical Forests," *Environmentalist* 8:3, pp. 1–20.
Nordhaus, W. 1992. "An Optimal Transition Path for Controlling Greenhouse Gases," *Science* 225, pp. 1315–1319.
Palmquist, R. 1991. "Hedonic Methods" in J. Braden and C. Kolstad (eds.), *Measuring the Demand for Environmental Quality*. Amsterdam: Elsevier Science Publishers.
Pearce, D. 1991. "Deforesting the Amazon: Toward an Economic Solution," *Ecodecision* 1, pp. 40–49.
Peck, S., and T. Teisberg. 1992. "CETA: A Model for Carbon Emissions Trajectory Assessment," *The Energy Journal* 13:1, pp. 15–77.
Peters, C., A. Gentry, and R. Mendelsohn. 1989. "Valuation of an Amazonian Rainforest," *Nature* 339, pp. 655–656.
Pichón, F., and R. Bilsborrow. 1992. "Land Use Systems, Deforestation, and Associated Demographic Factors in the Humid Tropics: Farm-Level Evidence from Ecuador," International Union for the Scientific Study of Popu-

lation Seminar on Population and Deforestation in the Humid Tropics, Campinas.

Ramos, H., and M. Acosta. 1991. "Impactos de la Apertura Comercial Regional en el Sector Agropecuario" (Technical Document No. 26), Instituto de Estrategias Agropecuarias, Quito.

Ramos, H., and L. Robison. 1990. "Credit and Credit Policies" in M. Whitaker and D. Colyer (eds.), *Agriculture and Economic Survival: The Role of Agriculture in Ecuador's Economic Development*. Boulder: Westview Press.

Rosenberry, B. 1990. "World Shrimp Farming: Can the Western Hemisphere Compete with the Eastern?" *Aquaculture Magazine* 16:5, pp. 60–64.

S.A. AGRER N.V. and Instituto Ecuatoriano de Recursos Hidráulicos (S.A. AGRER N.V./INERHI). 1982. "Proyecto El Pisque: Plan Maestro Quincenal," Quito.

Sanchez, R. 1992. "Estado Actual de los Manglares en el Ecuador" in *Teledetección: XV Aniversario de Creación de CLIRSEN*. Quito: Centro de Levantamientos Integrados de Recursos Naturales por Sensores Remotos.

Schmidt, R. 1990. "Sustainable Development of Tropical Moist Forests," Forestry Department, Food and Agriculture Organization of the United Nations, Rome.

Schneider, R., J. McKenna, C. Dejou, J. Butler, and R. Barrows. 1991. "Brazil: An Analysis of Environmental Problems in the Amazon" (Report to Latin America and Caribbean Region), International Bank for Reconstruction and Development, Washington.

Schodt, D. 1987. *Ecuador: An Andean Enigma*. Boulder: Westview Press.

Scobie, G., V. Jardine, and D. Greene. 1990. "The Importance of Trade and Exchange Rate Policies for Agriculture in Ecuador," *Food Policy* 16:1, pp. 34–47.

Scott, I., and N. Gaibor. 1992. "A Review of the Fishery for Shrimp Larvae in Ecuador: Biological, Economic, and Social Factors," Instituto Nacional de Pesca, Guayaquil.

Seligson, M. 1984. "Land Tenure Security, Minifundization, and Agrarian Development in Ecuador," U.S. Agency for International Development, Quito.

Sevilla-Larrea, R., and P. Pérez de Sevilla. 1985. "Los Plaguicidas en el Ecuador: Mas Allá de una Simple Advertencia," Fundación Natura, Quito.

Snedaker, S., J. Dickinson, M. Brown, and E. Lahmann. 1986. "Shrimp Pond Siting and Management Alternatives in Mangrove Ecosystems in Ecuador," U.S. Agency for International Development, Quito.

Solórzano, L. 1989. "Status of Coastal Water Quality in Ecuador" in S. Olsen and L. Arriaga (eds.), *A Sustainable Shrimp Industry for Ecuador*. Narragansett: University of Rhode Island Coastal Resources Center.

Soria, G. 1989. "El Hurto de Energía Eléctrica," presented at CEPE/INECEL Seminar on Petroleum, Energy, and the Environment, Quito.

Soria, J. 1990. "Agricultura y Medio Ambiente," Fundación Natura, Quito.

Southgate, D. 1990a. "The Causes of Land Degradation along 'Spontaneously' Expanding Agricultural Frontiers in the Third World," *Land Economics* 66:1, pp. 93–101.

Southgate, D. 1990b. "Development of Ecuador's Renewable Natural Resources" in M. Whitaker and D. Colyer (eds.), *Agriculture and Economic*

Survival: The Role of Agriculture in Ecuador's Economic Development. Boulder: Westview Press.

Southgate, D. 1991. "Tropical Deforestation and Agricultural Development in Latin America" (Discussion Paper No. 91–01), London Environmental Economics Centre, London.

Southgate, D. 1992a. "El Desarrollo Petrolero en Bosques Húmedos Tropicales: La Economía del Control de la Contaminación en el Oriente Ecuatoriano" (Technical Document No. 39), Instituto de Estrategias Agropecuarias, Quito.

Southgate, D. 1992b. "Petroleum Development in Tropical Rainforests: The Economics of Pollution Control in Eastern Ecuador," *Ecodecisión* 5, pp. 78–81.

Southgate, D., L. Chase, and M. Hanrahan. 1993. "Reversing Tropical Deforestation: Some Lessons from Northwestern Ecuador," Southern Forestry Economics Workshop, Durham.

Southgate, D., with M. Hanrahan, M. Bonifaz, M. Camacho, M. Carey, and L. Chase. 1992. "The Economics of Agricultural Land Clearing in Northwestern Ecuador," Instituto de Estrategias Agropecuarias, Quito.

Southgate, D., and R. Macke. 1989. "The Downstream Benefits of Soil Conservation in Third World Hydroelectric Watersheds," *Land Economics* 65:1, pp. 38–48.

Southgate, D., R. Stewart, V. Molinos, F. Guerrón, and B. Kernan. 1993. "Improving Incentives for Sustainable Forest Management: An Ecuadorian Case Study," Instituto de Estrategias Agropecuarias, Quito.

Subsecretaría Forestal y de Recursos Naturales (SUFOREN). 1991. *Diagnóstico: Plan de Acción Forestal, 1991–1995.* Quito: Ministerio de Agricultura y Ganadería.

Sutinen, J., J. Broadus, and W. Spurrier. 1989. "An Economic Analysis of Trends in the Shrimp Cultivation Industry in Ecuador" in S. Olsen and L. Arriaga (eds.), *A Sustainable Shrimp Industry for Ecuador.* Narragansett: University of Rhode Island Coastal Resources Center.

Sweeney, J. 1984. "The Response of Energy Demanded to Higher Prices," *American Economic Review Papers and Proceedings* 74:1, pp. 31–37.

Thia-Eng, C., and P. Kungvankij. 1989. "An Assessment of Shrimp Culture in Ecuador and Policy Strategy for Its Development and Mariculture Diversification," Coastal Resources Management Project, U.S. Agency for International Development, Guayaquil.

Thoumi, F. 1990. "The Hidden Logic of 'Irrational' Economic Policies in Ecuador," *Journal of Interamerican Studies and World Affairs* 32:2, pp. 43–68.

Tschirley, D., and H. Riley. 1990. "The Agricultural Marketing System" in M. Whitaker and D. Colyer (eds.), *Agriculture and Economic Survival: The Role of Agriculture in Ecuador's Economic Development.* Boulder: Westview Press.

Turner, R. 1989. "Factors Affecting the Relative Abundance of Shrimp in Ecuador" in S. Olsen and L. Arriaga (eds.), *A Sustainable Shrimp Industry for Ecuador.* Narragansett: University of Rhode Island Coastal Resources Center.

Twilley, R. 1989. "Impacts of Shrimp Mariculture Practices on the Ecology of Coastal Ecosystems in Ecuador" in S. Olsen and L. Arriaga (eds.), *A Sustain-*

able Shrimp Industry for Ecuador. Narragansett: University of Rhode Island Coastal Resources Center.

Valdés, A. 1986. "Impact of Trade and Macroeconomic Policies on Agricultural Growth: The South American Experience" in *Economic and Social Progress in Latin America.* Washington: Inter-American Development Bank.

Villalón, J., P. Maugle, and R. Laniado. 1989. "Present Status and Future Options for Improving the Efficiency of Shrimp Mariculture" in S. Olsen and L. Arriaga (eds.), *A Sustainable Shrimp Industry for Ecuador.* Narragansett: University of Rhode Island Coastal Resources Center.

Whitaker, M. 1990. "The Human Capital and Science Base" in M. Whitaker and D. Colyer (eds.), *Agriculture and Economic Survival: The Role of Agriculture in Ecuador's Economic Development.* Boulder: Westview Press.

Whitaker, M., and J. Alzamora. 1990a. "Irrigation and Agricultural Development" in M. Whitaker and D. Colyer (eds.), *Agriculture and Economic Survival: The Role of Agriculture in Ecuador's Economic Development.* Boulder: Westview Press.

Whitaker, M., and J. Alzamora. 1990b. "Production Agriculture: Nature and Characteristics" in M. Whitaker and D. Colyer (eds.), *Agriculture and Economic Survival: The Role of Agriculture in Ecuador's Economic Development.* Boulder: Westview Press.

Whitaker, M., D. Colyer, and J. Alzamora. 1990. *The Role of Agriculture in Ecuador's Economic Development: An Assessment of Ecuador's Agricultural Sector.* Quito: Instituto de Estrategias Agropecuarias.

Whitaker, M., and D. Greene. 1990. "Development Policy and Agriculture" in M. Whitaker and D. Colyer (eds.), *Agriculture and Economic Survival: The Role of Agriculture in Ecuador's Development.* Boulder: Westview Press.

Willardson, L., and B. Anderson. 1993. "An Experience in Irrigation Privatization," Fifteenth International Congress of the International Commission on Irrigation and Drainage, The Hague.

Wilson, E. (ed.). 1988. *Biodiversity.* Washington: National Academy Press.

World Commission on Environment and Development (WCED). 1987. *Our Common Future.* Oxford: Oxford University Press.

World Resources Institute (WRI). 1990. *World Resources 1990–91.* Washington.

Index

Agoyán Complex, 68–69, 75
agrarian reform, 25–26, 50, 52–53, 129–30. *See also* ecosystem destruction as a prerequisite for property rights; IERAC; property rights and resource development
agricultural chemical contamination. *See* water pollution
agricultural land clearing. *See* tropical deforestation
AID, vii–viii, 130
Argentina, 3, 118
Atlantic Richfield Company, 85, 121
Austria, 24
Azua Project, 128

Belize, 46
Bhutan, 24
biodiversity loss
    caused by tropical deforestation, 33, 40–41
    in the Galápagos, 101, 103–4, 106

Block 16, 83–88. *See also* CONOCO; Foreign Corrupt Practices Act; Huaorani Indians; Inter-American Commission on Human Rights; Maxus Oil Company; RAN; Sierra Club Legal Defense Fund; Yasuní National Park
BNF, 27
Bolivia, 13, 54, 119
Borja, Rodrigo, 19
Botswana, 24
Brazil, 3, 12, 15, 24, 36–37, 41, 89, 117, 119, 122–3
British Gas, 85
Brundtland Commission, 4

Carrizal-Chone Project, 60, 62–63
CEDEGE, 59, 62–63, 65, 75, 77, 128
CEPAR, 12
Charles Darwin Research Foundation and Station, 104–6
Chespi Project, 68–69

Chile, 9, 19, 24, 117–8, 120–1
China, 99
Churute Ecological Reserve, 96
CIP, 76
CLIRSEN, 91
coastal ecosystem degradation, 90–92. *See also* Churute Ecological Reserve; CLIRSEN; DGMM
  arising because of open-access status of public coastal lands, 96
  encouraged by low fees for shrimp pond permits, 24, 96–97, 119
  encouraged by property arrangements, 7, 91, 96–97, 100
  occurring because environmental externalities are ignored, 96–97
  possibly accelerated by lower shrimp prices, 99
  resulting from low productivity in shrimp mariculture, 7, 91, 97–99
  wild shrimp stocks reduced by, 94
cocoa boom, 34
Colombia, 12–13, 15, 28, 54, 115–6
communal tenure. *See* agrarian reform
CONOCO, 84–85, 121
Costa Rica, 24, 46, 115–6
CREA, 59
CRM, 59, 63
crop production, 8, 18, 27–28, 50, 115
  affected by income growth and public policies, 17–19, 30
  soil resource impacts of, 50
Cuyabeno Wildlife Reserve, 83
Czechoslovakia, 24

Darwin, Charles, 101, 103–5
Daule-Peripa Project, 60, 62–63, 68, 72, 75, 77, 94, 128
debt problems and economic austerity, 11, 16, 19, 120–1
  agricultural commodity demand affected by, 16

  sustainable development linkages to, 113–4, 120–3
DGMM, 96
DIGEMA, 82
DINAMA, 82–83
Dominican Republic, 128–9

Earth Summit, 118
economic liberalization, 9, 19, 118, 126. *See also* crop production; livestock production
  agricultural commodity demand increased because of, 17
  political opposition to, 118, 131–2
  renewable resource development affected by, 6, 11, 119, 125–6
  rural economy stimulated by, 19–20
ecosystem destruction as a prerequisite for property rights, 24–25, 127. *See also* coastal ecosystem degradation; IERAC; tropical deforestation
El Niño storms, 19, 95
EMBRAPA, 117
EPA, 95
externality problems. *See* market failure

Febres Cordero, León, 19
financial intermediation. *See also* BNF
  required for efficient resource management, 21, 27, 43, 130
  suppressed by weak property rights and governmental policies, 21, 26–27, 130
flooding problems, 75
Flores, Juan José. *See* Galápagos Islands
FONARYD, 61
Foreign Corrupt Practices Act, 84
fuel subsidies, 71, 97–98, 121, 130

Galápagos Islands. *See also* biodiversity loss; Charles Darwin Research Foundation and Station; Darwin, Charles; Galápagos National Park; INEFAN;

IUCN; migration; population growth; PRONAREG; UNESCO
growth of tourism in, 101, 105–8
history of human occupation of, 103
low fees paid by tourism operators in, 7, 24, 101–2, 107–9, 119, 127
scientific interest in, 104–5
unique flora and fauna of, 101–3
Galápagos National Park, 105–7
global warming, 33, 41–43
governmental claims on natural resources, 3, 21, 23–24, 59, 127. *See also* coastal ecosystem degradation; ecosystem destruction as a prerequisite for private property rights; Galápagos National Park; irrigation development; petroleum development; privatization of natural resources; tropical deforestation; underpricing of natural resources
traditional resource rights that conflict with, 44, 59
greenhouse effect. *See* global warming
Guatemala, 12, 15

Hayes, Randy. *See* RAN
Honduras, 12, 15, 98
Huaorani Indians, 84
human capital formation and sustainable development, 7, 22, 28–30, 89, 98, 123, 131
Hurtado, Osvaldo, 19, 119
Huxley, Julian. *See* UNESCO
hydroelectricity development. *See also* Agoyán Complex; Chespi Project; Daule-Peripa Project; INECEL; Pacific Corp; Paute Complex; San Francisco Project
capital spending on, 68
distribution of gains from, 71
electricity price subsidies and, 6, 70–71

impacts of inefficient operations and theft on, 68–70
impacts of poor watershed management on, 74–75

IDB, 74, 95
IDEA, vii–viii
IERAC, 25, 44, 96, 127, 129
import substitution and industrialization strategy of development. *See* macroeconomic policy distortions
income growth, 15–16. *See also* petroleum development
agricultural commodity demand growth caused by, 11, 16–17
Indonesia, 99, 115–6
INECEL, 59, 68, 70–71, 74, 78, 122
INEFAN, 36–37, 83, 96, 107–8, 127
INERHI, 57–67, 78, 128
INIAP, 65
Inter-American Commission on Human Rights, 84
irrigation development, 58–59, 61–64. *See also* Azua Project; Carrizal-Chone Project; CEDEGE; CREA; CRM; Daule-Peripa Project; FONARYD; INERHI; Jubones Project; Milagro Project; Montúfar Project; Patate Project; Pisque Project; Poza Honda Reservoir; PREDESUR; Puyango-Tumbez Project; Santa Elena Project; water pollution; water resources
benefits of, 65–67
budgets for, 61–65, 130
distribution of gains from, 67, 78, 119, 129
environmental consequences of, 48, 77
influence of technology transfer on the success of, 67, 77
privatization of irrigation systems, 128–9
public sector administration of, 59–61, 77–78

irrigation development (*Cont.*)
  water subsidies and, 6, 57, 62, 64–65, 119, 127, 130
IUCN, 104

Jubones Project, 60–61

labor market distortions, 22, 28–29
land markets. *See* governmental claims on natural resources; property rights and resource development
livestock production, 18–19, 49, 52, 54–55
  affected by income growth and government policies, 18–19, 54
  in deforested areas and soil resource impacts of, 34, 36, 49–50

macroeconomic policy distortions 15, 17, 22, 125–6. *See also* crop production; economic liberalization; livestock production
  agricultural commodity demand reduced because of, 16–17
  distributional consequences of, 118–9, 123
  renewable resource development affected by, 6–7, 17, 21–23, 53–54, 123
  rural economy depressed by, 17–19, 123, 125–6
MAG, 37–38, 45, 56
market failure, 4, 119, 124–5. *See also* biodiversity loss; coastal ecosystem degradation; global warming; overfishing of wild shrimp; petroleum development; soil erosion; tropical deforestation; water pollution
Maxus Oil Company, 84
MEM, 82–83, 85
Menem, Carlos, 118
Mexico, 9, 12, 15, 19, 53, 98–99, 118, 120–1, 130
migration
  from rural to urban areas, 10–11, 14–15, 55–56
  to agricultural frontiers, 10–11, 13–14, 34, 42–43, 84
  to the Galápagos, 106
  to the United States, 13
Milagro Project, 60, 65
military spending, 122
Montúfar Project, 60, 65

national patrimonies. *See* governmental claims on natural resources
natural resource scarcity, 115–7
  and environmental degradation, 7, 11, 20–22, 117
Norway, 24

overfishing of wild shrimp, 92–93
  arising because of open-access problems, 96
  caused by mariculture's low productivity, 7, 98
  economic consequences, 94

Pacific Corp, 70
Panama, 24
Papallacta Project, 69, 72–73
Paraguay, 12, 15, 46
Patate Project, 128
Paute Complex, 68–71, 74–75, 77
Peru, 12–13, 15, 28, 39, 54
Petrocanada, 85
Petroecuador, 82–86, 121–2
petroleum development. *See also* Atlantic Richfield Company; Block 16; British Gas; CONOCO; Cuyabeno Wildlife Reserve; DIGEMA; DINAMA; governmental claims on natural resources; INEFAN; Maxus Oil Company; MEM; Petrocanada; Petroecuador; SMA; Texaco; Trans-Ecuadorian Pipeline; Shushufindi gas plant
  cost of limiting pollution resulting from, 7, 86–88
  environmental impacts of, 9, 76
  impacts on national economy and public finances of, 3, 7,

15–16, 18–19, 71, 79, 87–88, 122
migration to agricultural frontiers associated with, 13–14, 34
potential benefits of increased private involvement in, 122
untapped reserves for, 121–2
weak regulation of pollution resulting from, 82–83, 119
Philippines, 99, 115–6, 126
Pisque Project, 60, 67
Ponce, Camilo. *See* Charles Darwin Research Foundation and Station
population growth, 5–6, 10–13, 114. *See also* CEPAR; migration
agricultural commodity demand growing because of, 10, 15–17
environmental degradation and, 4, 7, 42–43, 113, 115–7
in the Galápagos, 106
slowing because of recent declines in human fertility, 12–13, 114
potable water development, 107. *See also* Daule-Peripa Project; Papallacta Project; Poza Honda Reservoir; Santa Elena Project
capital spending on, 72–73
distribution of gains from, 72
governmental involvement in, 72
potable water subsidies and, 72–73
Poza Honda Reservoir, 69, 72, 77
PREDESUR, 59–60, 63, 65
privatization of natural resources, 121–2, 128–9. *See also* ecosystem destruction as a prerequisite for property rights; irrigation development
Project CARE, 56
PRONAMEC, 50
PRONAREG, 37–38, 103
property rights and resource development, 7, 20–21, 25–26, 46, 52–53, 56, 126–7, 129. *See also* agrarian reform; ecosystem destruction as a prerequisite for property rights; governmental claims on natural resources; privatization of natural resources; soil erosion; tropical deforestation
Puyango-Tumbez Project, 60, 63

RAN, 84–85
Rockefeller Foundation, 76

Salinas, Carlos, 118, 130
San Francisco Project, 68–69
Santa Elena Project, 60, 62–63
shrimp mariculture, 7, 18–19, 90, 115. *See also* coastal ecosystem destruction; overfishing of wild shrimp; water pollution
Shushufindi gas plant, 80
Sierra Club Legal Defense Fund, 84
SMA, 82–83
soil erosion, 37, 49–50. *See also* crop production; livestock production; Project CARE; PRONAMEC
agricultural production reduced because of, 4, 51
caused by declining commodity prices, 53–55
caused by policies that confine small farmers to fragile lands, 4, 22, 28–29, 50, 52
caused by policies that discourage the adoption of soil conservation measures, 6, 26, 52–53, 55
linked to rural-to-urban migration, 55–56
reduced by soil conservation projects, 51–52, 56
water resource development impaired by, 74–75
Spanish colonial era, 3, 34, 49–50
species extinction. *See* biodiversity loss
SPNG. *See* Galápagos National Park
SUFOREN. *See* INEFAN
Surinam, 46

Tanzania, 24
technological progress and sustainable development, 6–7, 20–22, 27–28, 30, 117–8, 131. *See also* coastal ecosystem degradation; INIAP; irrigation development; tropical deforestation
Texaco, 83
Thailand, 99, 115–6
timber production, 18–19, 39, 115. *See also* tropical deforestation
  affected by trade policies, 18–19, 44–46
Trans-Ecuadorian Pipeline, 80–81
tropical deforestation. *See also* biodiversity loss; global warming; migration; petroleum development
  agricultural benefits of, 37–38
  arising because of open-access status of public lands, 23–24, 43–44, 132
  caused by low productivity of agricultural economy, 6, 43, 117–8, 132
  caused by underpriced access to timber and forested lands, 6, 24, 44, 119, 127
  current rate of, 36–37
  encouraged by frontier property arrangements, 24, 44, 127
  history of, 34–36
  indigenous forest dwellers affected by, 33, 44
  occurring because environmental externalities are ignored, 42–43, 46, 124
  losses of timber and non-forest products associated with, 39–40

underpricing of natural resources, 24, 127. *See also* coastal ecosystem degradation; Galápagos Islands; hydroelectricity development; irrigation development; potable water development; tropical deforestation
UNESCO, 104
urbanization. *See* migration

water harvesting, 75
water pollution
  caused by shrimp mariculture, 93
  control techniques in newly developed oil fields, 84, 86
  impacts on shrimp production, 93–95
  regulation of oil industry water pollution, 82–83
  relationship to agricultural development, 76–77, 95
  resulting from petroleum development, 80–82
water resources, 58–59, 115. *See also* water harvesting; water pollution
  rural-versus-urban competition over, 72
World Bank, 75, 89, 114, 121, 130
WRI, 36

Yasuní National Park, 83

Zimbabwe, 115–6